T0377591

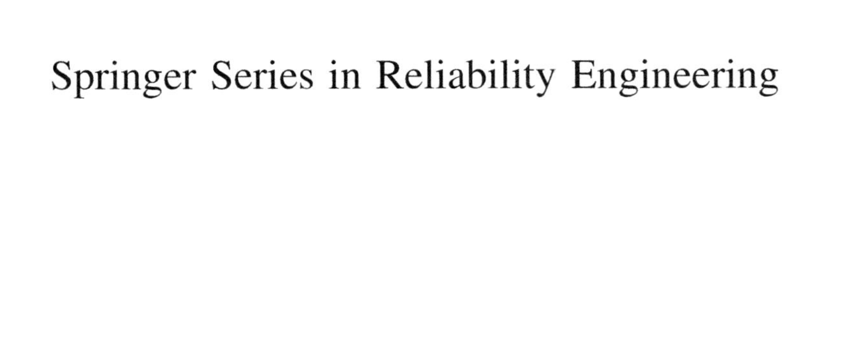

Springer Series in Reliability Engineering

For further volumes:
http://www.springer.com/series/6917

Ajit Kumar Verma · Srividya Ajit
Manoj Kumar

Dependability of Networked Computer-based Systems

Prof. Ajit Kumar Verma
Department of Electrical Engineering
Indian Institute of Technology Bombay
(IITB)
Powai, Mumbai 400076
India
e-mail: akvmanas@gmail.com

Dr. Manoj Kumar
System Engineering Section
Control Instrumentation Division
Bhabha Atomic Research Centre (BARC)
Trombay, Mumbai 400085
India
e-mail: kmanoj@barc.gov.in

Prof. Srividya Ajit
Department of Civil Engineering
Indian Institute of Technology Bombay
(IITB)
Powai, Mumbai 400076
India
e-mail: asvidya@civil.iitb.ac.in

ISSN 1614-7839

ISBN 978-0-85729-317-6 e-ISBN 978-0-85729-318-3

DOI 10.1007/978-0-85729-318-3

Springer London Dordrecht Heidelberg New York

British Library Cataloguing in Publication Data
A catalogue record for this book is available from the British Library

Cover design: eStudio Calamar, Berlin/Figueres

Printed on acid-free paper

Springer is part of Springer Science+Business Media (www.springer.com)

Dedicated to

Our **Parents**

Sri Kedar Prasad Verma & Smt Sushila Verma

Sri B.C. Khanapuri & Smt Vibhavati B. Khanapuri

Sri Gayanand Singh & Smt Droupdi Devi

Our **Gurus**

Bhagwan Sri Sathya Sai Baba

Paramhansa Swami Sathyananda Saraswati

Sri B. Jairaman & Smt Vijaya Jairaman

Dr. C.S. Rao & Smt Kasturi Rao

Our **Teachers**

Prof. A.S.R. Murthy (Reliability Engg., IIT Kharagpur)

Prof. M.A. Faruqi (Mechanical Engg., IIT Kharagpur)

Prof. N.C. Roy (Chemical Engg., IIT Kharagpur)

Foreword

A compelling requirement in today's context of the prevalent use of complex processes and systems in process and service industries, military applications, aerospace automotive and transportation, logistic, and other, is to maintain high *dependability* and *security*. The first casualty of integrated real time complex systems is dependability owing to an combinatorial like explosion of possible states, some with unacceptable probability levels and many unsafe states as well. The theoretical foundation for analytical treatment of such systems is crucial to their understanding, design and implementation. This book aims to explore the analysis, simulation and limitations in the implementation of such complex systems, addressing a multitude of issues and challenges in the application of computer-based systems in dependability and safety critical applications. In my opinion, this book is exceptional as it fulfills a long felt need of engineers, scholars, researchers and designers for a coherent, yet effective and efficient treatment of such systems and problems, built up with a conceptual hierarchy that starts from the basics.

The real-time systems also need to have an extremely important characteristic feature, timeliness. Most of the contemporary research work focuses on worst-case timing guarantees. A probabilistic *measure of timeliness* is the new buzzword in the dependability community. A highlight of this book is its emphasis on the probabilistic measures of dependability and timeliness. System designers usually face several problems while choosing an appropriate technology (in particular, a network) for a given project. The derivation of a timeliness hazard rate enables an integrated dependability modeling of the system which in turn helps the system designer in making decisions based on specific requirements and uniform measure(s). This book contains some detailed, well explained and intuitively appealing examples which the practitioner may find directly applicable in the analysis and solution of his or her problems.

I wish to congratulate the authors for their endeavors in bringing forth such a timely and insightful book on the dependability of networked computer-based systems. Their deep knowledge of the area, combined with vision concerning the present and future challenges, have led to this remarkable book. I am sure that this book will serve as an invaluable guide for scholars, researchers and practitioners interested and working in the field of critical applications where reliance on automation is indispensable.

October 2010

Academician Janusz Kacprzyk
Professor, Ph.D., D.Sc., Fellow of IEEE, IFSA
President of the Polish Society for Operational and Systems Research, Immediate Past President of IFSA (International Fuzzy Systems Association), Systems Research Institute, Polish Academy of Sciences,
Warsaw, Poland

Preface

This book is meant for research scholars, scientists and practitioners involved with the application of computer-based systems in critical applications. Ensuring dependability of systems used in critical applications is important due to the impact of their failures on human life, investment and environment. The individual aspects of system dependability—reliability, availability, safety, timeliness and security are the factors that determine application success. To answer the question on reliance on computers in critical applications, this book explores the integration of dependability attributes within practical, working systems. The book addresses the growing international concern for system dependability and reflects the important advances in understanding how dependability manifests in computer-based systems.

Probability theory, which began in the seventeenth century is now a well-established branch of mathematics and finds applications in various natural and social sciences, i.e. from weather prediction to predicting the risk of new medical treatments. The book begins with an elementary treatment of the basic definitions and theorems that form the foundation for the premise of this work. Detailed information on these can be found in the standard books on probability theory and stochastic theory, for a comprehensive appraisal. The mathematical techniques used have been kept as elementary as possible and Markov chains, DSPN models and Matlab code are given where relevant.

Chapter 1 begins with an introduction to the premise of this book, where dependability concepts are introduced. Chapter 2 provides the requisite foundation on the essentials of probability theory, followed by introduction to stochastic processes and models in Chap. 3. Various dependability models of computer-based systems are discussed in Chap. 4. Markov models for the systems considering safe failures, perfect and imperfect periodic proof tests, and demand rate have been derived. Analysis has been done to derive closed form solution for performance-based safety index and availability.

In Chap. 5, medium access control (MAC) protocol mechanisms of three candidate networks are presented in detail. The MAC mechanism is responsible for the access to the network medium, and hence effects the timing requirement of message transmission. A comparison of network parameters is also presented to provide an understanding of the various network protocols that can be used as primary guidelines for selecting a network solution for a given application.

Methods to probabilistically model network induced delay of two field bus networks, CAN, MIL-STD-1553B and Ethernet are proposed in Chap. 6. Hazard rates are derived from discrete time process for a fault tolerant networked computer system. Models are derived for the three dependability attributes—reliability, availability and safety, of NRT systems in Chap. 7. Timeliness hazard rate is modeled as reward rate.

We hope this book will be a very useful reference for practicing engineers and research community alike in the field of networked computer-based systems.

Mumbai, October 2010

Ajit Kumar Verma
Srividya Ajit
Manoj Kumar

Acknowledgments

We are indebted to Department of Electrical Engineering & Department of Civil Engineering, IIT Bombay and Control Instrumentation Division, BARC for their encouragement and support during the project.

Many of our friends, colleagues and students carefully went through drafts and suggested many changes changed improving the readability and correctness of the text. Many thanks to Shri U. Mahapatra, Shri G.P. Srivastava, Shri P.P. Marathe, Shri R.M. Suresh Babu, Shri M.K. Singh, Dr. Gopika Vinod, Prof. Vivek Agarwal, Prof. Varsha Apte and Prof. P.S.V. Nataraj for their suggestions. The help by publishing staff, especially of Mr. Claire, in timely preparation of the book is also appreciated.

Mumbai, October 2010

Ajit Kumar Verma
Srividya Ajit
Manoj Kumar

Contents

Acronyms

CAN	Controller area network
CSMA	Carrier sense multiple access
CTMC	Continuous time Markov chain
CCF	Common cause failure
DC	Diagnostic coverage
DD	Dangerous detected (failure category in IEC-61508)
DU	Dangerous undetected (failure category in IEC-61508)
DF	Dangerous failure (failure category in IEC-61508)
DEUC	Damage to EUC (or accident)
DSPN	Deterministic stochastic Petri net
EMI	Electromagnetic interference
EUC	Equipment under control or process plant
MBF	Multiple beta factor
MRM	Markov reward model
MTBD	Mean time between demands
MTTF	Mean time to failure
NLFS	Node level fault tolerance
NRT	Networked real-time
NCS	Networked control system
PES	Programmable electronic systems
PFD	Average probability of failure on demand
PFaD	Average probability of failure on actual demand
PFH	Probability of failure per hour
TDMA	Time division multiple access
TTA	Time triggered architecture
SF	Safe failure (failure category in IEC-61508)
QoP	Quality of performance
R_i	Response time of *i*th message
J_i	Queuing jitter of *i*th message
q_i	Queuing time for *i*th message
C_i	Worst-case transmission time of *i*th message

$d^{ij}(t)$	*pdf* of transmission delay from node i to node j
$D^{ij}(t)$	*CDF* of transmission delay from node i to node j
τ_x^i	Random variable denoting random time at node i for function x
λ^T	Timeliness hazard rate
$\mathrm{E}\ (\cdot)$	Expectation or mean operator
T_{proof}	Proof-test interval
mAv	Manifested availability
F_{DU}	Dangerous undetected state of the safety system
F_S	Safe failure state of the safety system
λ_{SF}	Hazard rate of a channel leading to SF
λ_{DD}	Hazard rate of a channel leading to DD
λ_{DU}	Hazard rate of a channel leading to DU
μ	Repair rate of a channel in FS
$\mu_p(t)$	Time dependent proof-test rate
λ_{arr}	Demand arrival rate (1/MTBD)
Δ	Probability redistribution matrix
Q	Infinitesimal generator matrix
Λ_{TT}	Transition matrix from transient state to transient states
Λ_{TA}	Transition matrix from transient state to absorbing states

Chapter 1
Introduction

Rapid advances in microelectronics and networking technology have lead to penetration of computers and networks into almost every aspect of our life. When these system are used in critical applications, such as, nuclear power plant, avionics, process plants and automobiles etc., failure of these systems could result in loss of huge investment, effort, life and damage to environment. In such case, dependability analysis becomes an important tool for decision making at all stages of system life-cycle – design, development, operation and phaseout. In fact for systems concerned with safety of critical facilities such as nuclear plants, demonstration of dependability through analysis is a mandatory requirement before system can be deployed.

Real-time systems refer to reactive computer-based systems, used in various control and on line processing applications requiring responses in real-time [1]. These computer systems are usually a part of a big system or network. Examples of such systems are fly-by-wire system of an aircraft, safety systems of a nuclear reactor, control system of vehicles (such as cars) and network routers to mention a few.

In real-time systems, missing deadline is as dangerous as producing incorrect response (i.e. *value*). So, real-time system failure has two causes, i) *value*, and ii) failing to produce response at correct time (i.e. *timeliness*). Most dependability models do not consider timeliness explicitly. They use worst-case guarantee to ensure timeliness. Worst-case analysis is a deterministic method of analysis, which considers worst-case scenarios irrespective of their likelihood. This is in contrast to component failures, where probabilistic methods are used for analysis. Two separate analysis for two types of failures pose difficulty in decision making. A probabilistic measure for timeliness and its incorporation in dependability models could solve this problem by providing a single dependability measure.

In this book, a method has been outlined to estimate response-time distribution (mainly for CAN and MIL-STD-1553B networks) and timeliness hazard rate.

A. K. Verma et al., *Dependability of Networked Computer-based Systems*,
Springer Series in Reliability Engineering, DOI: 10.1007/978-0-85729-318-3_1,

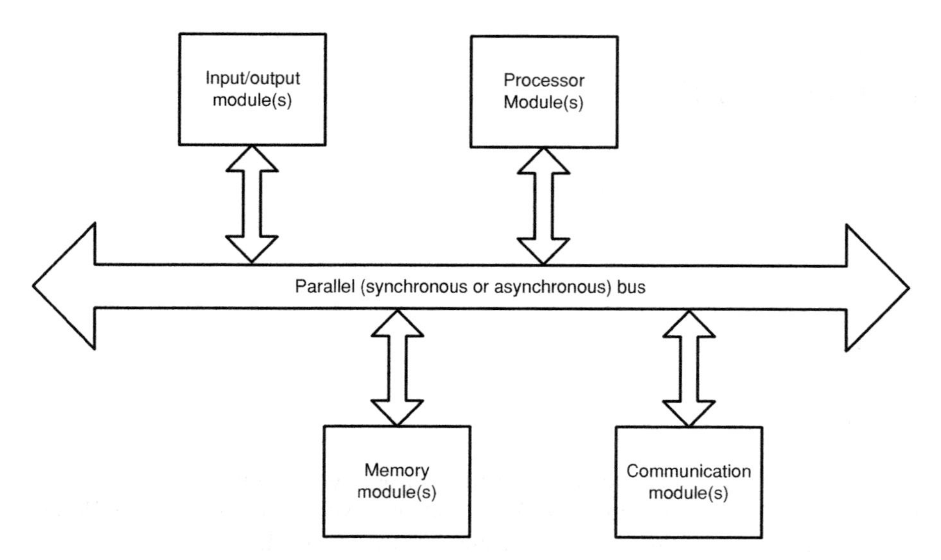

Fig. 1.1 A typical diagram of traditional real-time systems

Dependability models have been developed for networked real-time (NRT) systems, incorporating timeliness failures along with hardware failures.

Markov process and Markov regenerative process based model are used for dependability modeling. For higher level modeling and automated solution, DSPN based software tool- TimeNET can be used. For response-time distribution, basic probability theory operator such as, addition, minimum, maximum etc. are used.

Use of timeliness failure in dependability analysis is an evolving concept. In related works, task arrival is assumed to follow random arrival (Poisson). This basic concept has been extended to systems with periodic/deterministic tasks arrivals with asynchronous phase relation. The advantage of the proposed dependability model is it makes it possible to compare various designs with a single dependability measure considering both failures – value and timeliness.

1.1 Evolution of computer based systems

A typical diagram of a traditional real-time system is shown in Figure 1.1. The parallel bus could be synchronous (PCI) or asynchronous (EISA, VME, CAMAC, etc). To acquire physical inputs and generate physical outputs, input-output modules are used. Memory module(s) are used to store data and communication module(s) are used for communication with other systems. Processor module usually contain the computer (or processor) and performs the required logical & arithmetic operations as per the stored software.

Traditional real-time systems have a point-to-point communication, i.e. dedicated wires connecting computer with input points (or sensors) and output points

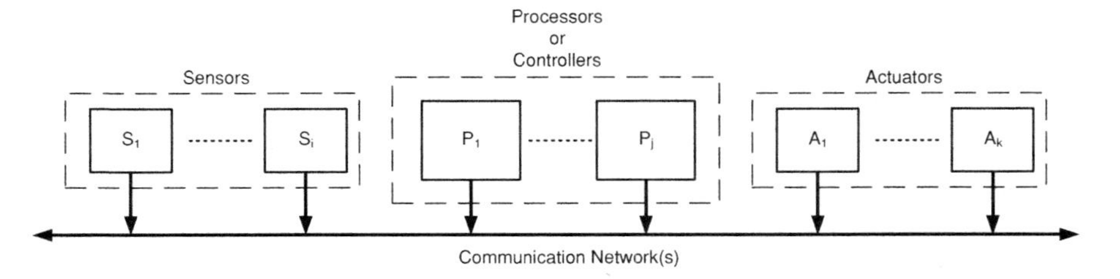

Fig. 1.2 A typical diagram of networked real-time systems [39, 40]

(or actuators) [2–7]. Disadvantages of this approach [2, 3] includes lack of following:

1. Modularity
2. Decentralization of control
3. Integrated diagnosis
4. Quick and easy maintenance
5. Low cost

A schematic of a networked real-time system is shown in Figure 1.2. Here system is composed of nodes interconnected through one or more shared communication network. The nodes are categorized in three groups; sensor, processor and actuators.

Networked real-time system is one which has different nodes carrying out different aspects of system's function and uses network to interact.

Networked system removes most of the limitations and offers cost effectiveness due to reduced cost of following:

1. Installation
2. Reconfiguration
3. Maintenances

The distributed nature of networked systems also introduces some problems, namely:

1. Network induced time varying delay [8]
2. Control system stability [9]
3. QoP [8]

1.2 Application areas: safety-critical, life-critical

Main characteristics of real-time systems that distinguishes them from others is that the correctness of the system depends on *value* as well as *time* of the response. Real-time systems that used in critical applications such as, nuclear, avionics and automobiles etc. Their failure can cause damage to huge investment, effort, life and/or environment. Based on the function(s) and extent of failure, these real-time systems are categorized in the following three types:

1. **Safety-critical systems**: Systems required to ensure safety of EUC (equipment under control), people and environment. Examples include, shutdown system of nuclear reactor, digital flight control computer of aircraft etc.
2. **Mission-critical systems**: Systems whose failure results in failure of mission. For example, control & coding unit (CCU) of an avionic system, navigation system of an spacecraft etc.
3. **Economically-critical systems**: Systems whose failure result in unavailability of EUC, causing massive loss of revenue. For example, Reactor control system of nuclear power plant.

Dependability attributes [10] for different kind of systems are different. For safety-critical systems, the dependability attribute, *safety* [11] is of concern. *Reliability* is the appropriate dependability measure for mission-critical systems [12–14]. Similarly, for economically-critical systems the dependability measure *availability* is of importance [15].

Extensive literature exists on dependability modeling of programmable electronic systems and/or real-time systems, i) Safety [11, 16–21], ii) Reliability [7, 13, 22–27], and iii) Availability [13–15]. Reliability models for soft real-time systems are discussed in [28–30]. Networked real-time system like real-time systems may fail due to value or timeliness.

1.3 A review of system failures

With rapid development in electronics, computer and communication technology, new technology based system has penetrated into almost every aspect of our society. These development coupled with improvement in productivity, yield and efficiency has resulted in cost effective deployment of these systems. These computer and communication systems have become the underpinning of many national infrastructures such as telecommunications, transport, water supply, electric power grid, banking and financial services, governmental services, and national defense systems, etc., which are vital to all crucial economic and social functions.

Besides the huge benefits obtained from networked systems, there exist remarkable risks and threats when a failure occurs. Some major network failure that have occurred during the past two decades are listed below:

- Outage of ARPAnet - the precursor of internet - for 4 hours on 27 October 1980. The cause of the problem include the lack of parity checking in router memory.
- 9-hour nationwide long distance service outage due to software bug in signaling system. It incurred about $60 million loss in revenue and a substantial damage to AT&T's reputation as a reliable long distance provider.

- Network outage for three day in a Medical center in Massachusetts in Nov. 2002. The network outage interrupted the hospital's Web-based applications such as email, patient monitoring, clinical records management, and prescription ordering. The cause of the problem was found to be a medical software producing excessive traffic and a faulty configuration in structure propagating the overload throughout the whole network.

From the above incidents it is clear that a networked system faces threats from three sources:

1. The system may have faults occurring in its hardware and faults present in its software subsystem.
2. A networked system tend to be fragile under adverse environmental conditions such as storms, earthquakes, flooding, fires, cable cut ands and power outages.
3. The most troubling aspect of present networked systems' weakness may come from malicious attacks.

The first problem has been tried with the help of fault tolerant techniques, fault diagnostics and fault removal. Today's hardware systems could commonly reduce the failure rate to a negligible values so that hardware faults are not a significant contributor to a system failures. On the other hand, software reliability and quality is far from satisfactory with its rapid growth in size and complexity. Software faults accounts for a substantial part for a system failures, and the situation is usually worsened by strict time constraints on functionality delivery in a competitive business environment.

Adequate redundancy and diversity seems to be the key for second problem. Diversity is important as common cause failure may defy redundancy, i.e. all redundant system might fail together.

The Internet has created an open universe to all people and organizations around the globe to interact with each other. However, this openness and the accessibility to a huge volume of information and computing resources may be taken advantage of by hackers, terrorists, and hostile states to launch attacks. Malicious attacks are evolving along with the development of the Internet itself. The same virtue of the Internet design that provides flexibility and productivity also makes the attacks difficult to predict, identify, and defend against.

1.4 Example: Comparison of system reliability

Let's consider a system with following reliability requirement, the system reliability for a mission time of 10,000 hrs with repair, shall be 0.9. The system implementation has three options, i) an analog, ii) a processor-based, and iii) a networked based.

Analog implementation uses analog components. Analog systems do not have sharing of resources, so for each function a dedicated resource is available. Also, they have limited fault-diagnosis and fault-coverage. Processor-based

Table 1.1 Reliabilities values

	R(10000)
Analog system	0.91
Process-based system	0.93
Network-based system	0.97

Table 1.2 Reliabilities values with probability of deadline missing

	R(10000)
Analog system	0.91
Process-based system	$0.93 \times 0.99 = 0.9207$
Network-based system	$0.97 \times 0.95 = 0.9215$

implementation on the other hand are quite simple, as most of the complex functions are implemented on a single processor. They have detailed fault-diagnosis and very good coverage. They can be easily interfaced with industry standard communication link, through which fault information may be communicated, displayed, stored and a detailed report can be generated. All this helps in a lower time to repair-time. Networked based system has all advantages of processor-based system, with much more modularity, i.e. further reduced repair-time and cost.

Reliability values considering hardware failures and repair-time for the three options are given in Table 1.1.

From the table it seems that network-based system gives better reliability. But the question is whether all failures mechanisms have been considered for processor and network-based system. These system may fail due to missing of time deadline. This problem of missing deadline is manageable to some extent in processor-based system but in networked-based system it is really a challenge. So, Table 1.1 is of limited use in making decision of which system will give better reliability - failure-free operation - in application.

Let probability of missing deadline for the mission time for processor and network-based system is given as 0.95 and 0.9, respectively, by some means. Now, the system reliability considering deadline miss is given in Table 1.2.

From Table 1.2, it is clear that all three systems provides similar level of reliability. This makes the comparison process uniform and simple.

The book provide methods which enable getting us the probability of deadline miss and their incorporation in dependability models.

1.5 Dependability

1.5.1 Basic concepts

Dependability is a collective term used to describe the ability to deliver service that can justifiably be trusted [10]. The service delivered by a system is its behavior as it is perceived by its user(s); a user is another system (physical, human) that interacts with the former at the service-interface. The function of a system is what the system is intended for, and is described by the system specification.

As per Laprie et al [10] concept of dependability consists of three parts: the threat to, the attributes of, and the means by which dependability is attained.

1.5.1.1 Threats

The threats are mainly three, failure, error and fault [10]. A system failure is an event that occurs when the delivered service deviates from correct service. A failure is a transition from correct service to incorrect service. An error is that part of system state that may cause a subsequent failure: a failure occurs when an error reaches the service interface and alters the service. A fault is the adjudged or hypothesized cause of an error. A fault is active when it produces an error, otherwise it is dormant.

1.5.1.2 Attributes

Dependability is an integrative concept that encompasses the following attributes:

- *Availability*: readiness for correct service [10, 31]
- *Reliability*: continuity of correct service [10, 31]
- *Safety*: absence of catastrophic consequences on the user(s) and environment [10, 32, 11]
- *Confidentiality*: absence of unauthorized disclosure of information [10, 31, 33, 34]
- *Integrity*: absence of improper system state alteration [10, 31, 33, 34]
- *Maintainability*: ability to undergo repairs and modifications [10, 31]

1.5.1.3 Means

Traditionally combination of following means is used for development of dependable computing system [10, 31]:

- *Fault prevention*: means to prevent the occurrence or introduction of faults
- *Fault tolerance*: means to deliver correct service in the presence of faults
- *Fault removal*: means to reduce the number or severity of faults

Another means to improve dependability has been proposed by Laprie et al [10], it is fault forecasting.

1.5.2 Basic Definitions and Terminology

Fault tolerance is the ability of a system to continue to perform its tasks after occurrence of faults. The fault-tolerance requires fault detection, fault containment, fault location, fault recovery and/or fault masking [12]. The definition of these terms is as follows:

- Fault detection is the process of recognizing that a fault has occurred.
- Fault containment is the process of isolating a fault and preventing the effects of that from propagating throughout the system.
- Fault location is the process of determining where a fault has occurred so that an appropriate recovery can be implemented.
- Fault recovery is the process of remaining operational or regaining operational status via reconfiguration even in the presence of fault.
- Fault masking is the process that prevents faults in a system from introducing errors into the informational structure of that system. A system employing fault masking achieves fault tolerance by "hiding" faults. Such systems do not require that fault be detected before it can be tolerated, but it is required that the fault be contained.

Systems that do not use fault masking require fault detection, fault location and fault recovery to achieve fault tolerance. Redundancy is essential for achieving fault tolerance. Redundancy is simply addition of information, resources, or time beyond what is needed for normal system operation. The redundancy can take any of the following form:

1. Hardware Redundancy is the addition of extra hardware, usually for the purpose of either detecting or tolerating faults.
2. Software Redundancy is the addition of extra software, beyond what is needed to perform a given function, to detect and possibly tolerate faults.
3. Information Redundancy is the addition of extra information beyond what is needed to implement a given function.
4. Time redundancy uses additional time to perform the functions of a system such that fault detection and often fault tolerance can be achieved.

1.5.2.1 Hardware Redundancy

The three basic form of redundancy:

1. Passive techniques use the concept of fault masking to hide the occurrence of faults and prevent the fault from resulting in errors. Examples of passive redundancy are Triple Modular Redundancy (TMR), N-Modular redundancy etc, where majority voting or median of the module outputs are taking to decide the final output and mask the fault of module(s).

2. Active techniques also referred as dynamic method, achieves fault tolerance by detecting the existence of faults and performing some action to remove the faulty hardware from the system. Examples are Duplication with comparison, standby sparing (cold or hot) etc, where faulty module is detected and control is transferred to the healthy one.
3. Hybrid techniques combine the attractive features of both the passive and active approaches. Fault masking is used in hybrid systems to prevent result from being generated. Fault detection, fault location, and fault recovery are also used in the hybrid approaches to improve fault tolerance by removing faulty hardware and replacing it with spares. Examples are N-Modular redundancy with spare, Self-purging redundancy, sift-out modular redundancy etc, which uses fault masking as well as reconfiguration.

1.5.2.2 Information Redundancy

Additional information is stored with the data in following forms:

1. Parity
2. m-of-n codes
3. duplication codes
4. checksums
5. cyclic codes
6. arithmetic codes, etc.

1.5.2.3 Time Redundancy

Time redundancy methods attempt to reduce the amount of extra hardware at the expense of using additional time. As it is clear, above two methods require use of extra hardware. So time redundancy becomes important in applications where more hardware cannot be put, but extra time can be provided using devices of higher speed.

1. Transient Fault Detection The basic concept of the time redundancy is the repetition of computations in ways that allow faults to be detected. In transient fault detection the concept is to perform the same computation two or more times and compare the results to determine if a discrepancy exists. If an error is detected, the computations can be performed again to see if the disagreement remains or disappears. Such approaches are often good for detecting errors resulting from transient faults, but they cannot protect against errors resulting from permanent faults.
2. Permanent Fault Detection One of the biggest potentials of time redundancy is the ability to detect permanent faults while using a minimum of extra hardware. The approaches for this are as follows:

   a. Alternating Logic
   b. Recomputing with Shifted Operands

c. Recomputing with Swapped Operands
d. Recomputing with Duplication with Comparison

3. Recomputation for Error Correction Time redundancy approaches can also provide error correction of logical operations if the computations are repeated three or more times. Limitation is this approach cannot work for arithmetic operations because the adjacent bits are not independent.

1.5.2.4 Software Redundancy

Software redundancy may come in many forms starting from few lines of extra code to complete replica of program. There could be few lines of code to check the magnitude of a signal or as a small routine used to periodically test a memory by writing and reading specific locations. The major software redundancy techniques are as follows:

1. Consistency check uses a priori knowledge about the characteristics of information to verify the correctness of that information.
2. Capability checks are performed to verify that a system possesses the capability expected.
3. N-Version Programming: Software does not break as hardware do, but instead software faults are the result of incorrect software designs or coding mistakes. Therefore, any technique that detects faults in software must detect design flaws. A simple duplication and comparison procedure will not detect software faults if the duplicated software modules are identical, because the design mistakes will appear in both modules.

The concept of N-version programming allows certain design flaws in software module to be detected by comparing design, code and results of N-versions of same software.

From the above discussion it is clear that use of time and software redundancy and reconfigurable hardware redundancy reduces the total hardware for achieving fault-tolerance. In case of distributed systems, processing units called nodes, are distributed and communicate through communication channels. Processing a task/job requires multiple nodes. System can be made fault-tolerant by transferring its processing to a different nodes, in case of error. Here system reconfigure itself and make use of time redundancy, as the new node has to process the tasks/jobs of the failed node in addition to its own.

1.6 Motivation

Real-time system used in critical applications are designed in accordance with stringent guidelines and codes [35–38]. They are fault-tolerant at hardware and protocol levels through redundancy and indulgent protocols [38], respectively.

Software application residing in these systems, has deterministic execution cycle [34, 37]. In literature, there is lack of probabilistic models for timeliness failures.

In literature, for real-time systems dependability attribute – reliability – is mainly considered. While different dependability attributes may be required based on the criticality of system failures, such as safety and availability.

A unified dependability model for NRT systems considering factors affecting *value* and *timeliness* will provide a rational platform to assess the dependability of NRT systems. These limitations are inspiration for this book.

1.7 Summary

Real-time systems has one additional failure mechanism, i.e. failure due to deadline miss. When networked systems are used in real-time application this failure mechanism becomes much more important. For ensuring timeliness, mostly worst-case response-time guarantees are used, which are deterministic. Systems with different timing characteristic cannot be compared based on a given dependability measure, as dependability models do not consider timeliness. The chapter points out this limitation with the help of an example. A refresher of means to achieve dependability is given for the sake of completeness.

References

1. Buttazzo GC (2003) Hard Real-time Computing Systems - Predictable Scheduling Algorithms and Applications. Springer
2. Lian F-L (2001) Analysis, Design, Modeling, and Control of Networked Control Systems. PhD thesis, University of Michigan
3. Lian F-L, Moyne J, Tilbury D (2001) Performance evaluation of control networks: Ethernet, controlnet, and devicenet. IEEE Control System Magazine 21:66–83
4. Lian F-L, Moyne J, Tilbury D (2002) Network design consideration for distributed control systems. IEEE Transaction on Control System Technology 10:297–307
5. Nilsson J (1998) Real-Time Control Systems with Delays. PhD thesis Lund Institute of Technology, Sweden
6. Nilsson J, Bernhardsson B, Wittenmark B (1998) Stochastic analysis and control of real-time systems with random time delays. Automatica 34:57–64
7. Philippi S (2003) Analysis of fault tolerant and reliability in distributed real-time system architectures. Reliability Engineering and System Safety 82:195–206
8. Yepez J, Marti P, Fuertes JM (2002) Control loop performance analysis over networked control systems. In: Proceedings of IECON2002
9. Zhang W, Branicky MS, Phillips SM (2001) Stability of networked control systems. IEEE Control System Magazine, p 84–99
10. Avizienis A, Laprie J-C, Randell B (2000) Fundamental concepts of dependability. In: Proc. of 3rd Information Survivability Workshop, p 7–11
11. IEC 61508: Functional safety of electric/electronic/programmable electronic safety-related systems, Parts 0-7; Oct. 1998-May (2000)

12. Johnson BW (1989) Design and Analysis of Fault-Tolerant Digital Systems. Addison Wesley Publishing Company
13. Mishra KB (1992) Reliability Analysis and Prediction. Elsevier
14. Trivedi KS (1982) Probability and Statistics with Reliability, Queueing, and Computer Science Applications. Prentice-Hall, Englewood Cliffs New Jersey
15. Varsha Mainkar. Availability analysis of transaction processing systems based on user perceived performance. In: Proceedings of 16th Symposium on Reliable Distributed Systems, Durham, NC, Oct. 1997.
16. Zhang T, Long W, Sato Y (2003). Availability of systems with self-diagnostics components-applying markov model to IEC 61508-6. Reliability Engineering and System Safety 80:133–141
17. Bukowski JV (2001) Modeling and analyzing the effects of periodic inspection on the performance of safety-critical systems. IEEE Transaction Reliability 50(3):321–329
18. Choi CY, Johnson BW, Profeta III JA (1997) Safety issues in the comparative analysis of dependable architectures.IEEE Transaction Reliability 46(3):316–322
19. Summers A (2000) Viewpoint on ISA TR 84.0.02-simplified methods and fault tree analysis.ISA Transaction 39(2):125–131
20. Bukowski JV (2005) A comparison of techniques for computing PFD average. In: RAMS 2005 590–595
21. Goble WM, Bukowski JV (2001) Extending IEC 61508 reliability evaluation techniques to include common circuit designs used in industrial safety systems. In: Proc. of Annual Reliability and Maintainability Symposium 339–343
22. Khobare SK, Shrikhande SV, Chandra U, Govidarajan G (1998) Reliability analysis of micro computer modules and computer based control systems important to safety of nuclear power plants. Reliability Engineering and System Safety 59(2):253–258
23. Jogesh Muppala, Gianfranco Ciardo, Trivedi KS (1994). Stochastic reward nets for reliability prediction. Communications in Reliability, Maintainability and Serviceability 1(2):9–20
24. Kim H, Shin KG (1997) Reliability modeling of real-time systems with deadline information. In: Proc. of IEEE Aerospace application Conference 511–523
25. Kim H, White AL, Shin KG (1998) Reliability modeling of hard real-time systems. In: Proceedings of 28th Int. Symp. on Fault Tolerant Computing 304–313
26. Tomek L, Mainkar V, Geist RM, Trivedi KS (1994) Reliability modeling of life-critical, real-time systems. Proceedings of the IEEE 82:108–121
27. Lindgren M, Hansson H, Norstrom C, Punnekkat S (2000) Deriving reliability estimates of distributed real-time systems by simulation.In: Proceeding of 7th International Conference on Real-time Computing System and Applications 279–286
28. Mainkar V, Trivedi KS (1994) Transient analysis of real-time systems using deterministic and stochastic petri nets. In: Int'l Workshop on Quality of Communication-Based Systems
29. Mainkar V, Trivedi KS (1995) Transient analysis of real-time systems with soft deadlines. In: Quality of communication based systems
30. Muppala JK, Trivedi KS Real-time systems performance in the presence of failures. IEEE Computer Magazine 37–47 May 1991.
31. Avizienis A, Laprie J-C, Randell B, Landwehr C (2004) Basic concepts and taxonomy of dependable and secure computing. IEEE Transaction Dependable and Secure Computing 1(1):11–33
32. Atoosa Thunem P-J (2005). Security Research from a Multi-disciplinary and Multi-sectoral Perspective. Lecture Notes in Computer Science (LNCS 3688). Springer Berlin / Heidelberg 381–389
33. Ross J.Anderson (2001) Security Engineering: A Guide to Building Dependable Distributed Systems. Wiley Computer Publishing, USA
34. MIL-STD-1553B: Aircraft internal time division command/response multiplex data bus, 30 April 1975.
35. AERB/SG/D-25: Computer based systems of pressurised heavy water reactor, 2001.

36. Safety guide NS-G-1.3 Instrumentation and control systems important to safety in nuclear power plants, 2002.
37. IEC 60880-2.0: Nuclear power plants - instrumentation and control systems important to safety - software aspects for computer-based systems performing category a functions, 2006.
38. Keidar I, Shraer A (2007) How to choose a timing model? In: Proc. 37th Annual IEEE/IFIP Int. Conf. on Dependable Systems and Networks (DSN'07)
39. Yang H, Sikdar B (2007) Control loop performance analysis over networked control systems. In: Proceedings of ICC 2007 241–246
40. Yang TC Networked control systems: a brief survey. IEE Proc.-Control Theory Applications 153(4):403–412, July 2006.

Chapter 2
Probability Theory

Probability theory deals with the study of events whose precise occurrence cannot be predicted in advance. These kind of events are also termed as random events. For example, a toss of a coin, the result may be either HEAD or TAIL. The precise result cannot be predicted in advance, hence the event—tossing a coin—is an example of random event.

Probability theory is usually discussed in terms of experiments and possible outcomes of the experiments.

2.1 Probability Models

Probability is a positive measure associated with each simple event. From a strict mathematical point of view it is difficult to define the concept of probability. A relative frequency approach, also called the posteriori approach is usually used to define probability.

In classical probability theory, all sample spaces are assumed to be finite, and each sample point is considered to occur with equal frequency. The definition of the probability P of an event A is described by relative frequency by which A occurs:

$$P(A) = \frac{h}{n} \tag{2.1}$$

where h is the number of sample points in A and n is the total number of sample points. This definition is also called probability definition based on relative frequency.

Probability theory is based on the concepts of sets theory, sample space, events and algebra of events. Before proceeding ahead, these will be briefly reviewed.

A. K. Verma et al., *Dependability of Networked Computer-based Systems*,
Springer Series in Reliability Engineering, DOI: 10.1007/978-0-85729-318-3_2,

2.2 Sample Space, Events and Algebra of Events

Probability theory is study of random experiments. The real life experiments may consist of the simple process of noting whether a component is functioning properly or has failed; measuring the response time of a system; queueing time at a service station. The result may consist of simple observations such as 'yes' or 'no' period of time etc. These are called outcomes of the experiment.

The totality of possible outcomes of a random experiment is called *sample space* of the experiment and it will be denoted by letter 'S'.

The sample space is not always determined by experiment but also by the purpose for which the experiment is carried out.

It is useful to think of the outcomes of an experiment, the elements of the sample space, as points in a space of one or more dimensions. For example, if an experiment consists of examining the state of a single component, it may be functioning correctly, or it may have failed. The sample space can be denoted as one-dimension. If a system consists of 2 components there are four possible outcomes and it can be denoted as two-dimensional sample space. In general, if a system has n components with 2 states, there are 2^n possible outcomes, each of which can be regarded as a point in n-dimensional space.

The sample space is conventionally classified according to the number of elements they contain. If the set of all possible outcomes of the experiment is finite, then the associated sample space is a finite sample space. Finite sample space is also referred as countable or a discrete sample space.

Measurement of time—in response-time, queueing time, time till failure—would have an entire interval of real numbers as possible values. Since the interval of real number cannot be enumerated, they cannot be put into one-to-one correspondence with natural numbers—such a sample space is said to be uncountable or non denumerable. If the elements of a sample space constitute a continuum, such as all the points of a line, all the points on a line segment, all the points in a plane, the sample space is said to be continuous.

A collection or subset of sample points is called an *event*. Means, any statement of conditions that defines this subset is called an event. The set of all experimental outcomes (sample points) for which the statement is true defines the subset of the sample space corresponding to the event. A single performance of the experiment is known as trial. The entire sample space is an event called the universal event, and so is the empty set called the null or impossible event. In case of continuous sample space, consider an experiment of observing the time to failure of a component. The sample space, in this case may be thought of as the set of all non-negative real numbers, or the interval $[0,\infty) = \{t|0 \le t < \infty\}$.

Consider an example of a computer system with five identical tape drives. One possible random experiment consists of checking the system to see how many tape drives are currently available. Each of the tape drive is in one of two states: busy (labeled 0) and available (labeled 1). An outcome of the experiment (a point in sample space) can be denoted by a 5-tuple of 0's and 1's. A 0 in position i of the

5-tuple indicates that tape drive i is busy and a 1 indicates that it is available. The sample space S has $2^5 = 32$ sample points.

A set is a collection of well defined objects. In general a set is defined by capital letter such as A, B, C, etc. and an element of the set by a lower case letter such as a, b, c, etc.

Set theory is a established branch of mathematics. It has a number of operations, operators and theorems. The basic operators are union and intersection. Some of the theorems are given below:

1. Idempotent laws:

$$A \cup A = A \quad A \cap A = A \tag{2.2}$$

2. Commutative laws:

$$A \cup B = B \cup A \quad A \cap B = B \cap A \tag{2.3}$$

3. Associative laws:

$$\begin{aligned} A \cup (B \cup C) &= (A \cup B) \cup C = A \cup B \cup C \\ A \cap (B \cap C) &= (A \cap B) \cap C = A \cap B \cap C \end{aligned} \tag{2.4}$$

4. Distributive laws:

$$\begin{aligned} A \cap (B \cup C) &= (A \cap B) \cup (A \cap C) \\ A \cup (B \cap C) &= (A \cup B) \cap (A \cup C) \end{aligned} \tag{2.5}$$

5. Identity laws:

$$\begin{aligned} A \cup \phi = A &\quad A \cap \phi = \phi \\ A \cup U = U &\quad A \cap U = A \end{aligned} \tag{2.6}$$

6. De Morgan's laws:

$$\begin{aligned} \overline{(A \cup B)} &= \overline{A} \cap \overline{B} \\ \overline{(A \cap B)} &= \overline{A} \cup \overline{B} \end{aligned} \tag{2.7}$$

7. Complement laws:

$$\begin{aligned} A \cup \overline{A} = U & \qquad\qquad A \cap \overline{A} = \phi \\ \overline{(\overline{A})} = A \quad & \overline{U} = \phi \quad \overline{\phi} = U \end{aligned} \tag{2.8}$$

8. For any sets A and B:

$$A = (A \cap B) \cup (A \cap \overline{B}) \tag{2.9}$$

A set of a list of all possible outcomes of an experiment is called sample space. The individual outcome is called sample point. For example, in an experiment of throwing a dice, the sample space is set of $\{1, 2, 3, 4, 5, 6\}$.

If the sample space has a finite number of points it is called a finite sample space. Further, if it has many points as there are natural numbers it is called a countable finite sample space or discrete sample space. If it has many points as there are points in some interval it is called a non-countable infinite sample space or a continuous sample space.

An event is a subset of sample space, i.e. it is a set of possible outcomes. An event which consists of one sample point is called a simple event.

2.3 Conditional Probability

Condition probability deals with the relation or dependence between two or more events. The kind of questions dealt with are '*the probability that one event occurs under the condition that another event has occurred*'.

Consider an experiment, if it is known that an event B has already occurred, then the probability that the event A has also occurred is known as the *conditional probability*. This is denoted by $P(A|B)$, the conditional probability of A given B, and it is defined by

$$P(A|B) = \frac{P(A \cap B)}{P(B)} \tag{2.10}$$

2.4 Independence of Events

Let there are two events A and B. It is possible for the probability of an event A to decrease, remain the same, or increase given that event B has occurred. If the probability of the occurrence of an event A does not change whether or not event B has occurred, we are likely to conclude that two events are independent if and only if:

$$P(A|B) = P(A) \tag{2.11}$$

From the definition of conditional probability, we have [provided $P(A) \neq 0$ and $P(B) \neq 0$]:

$$P(A \cap B) = P(A)P(B|A) = P(B)P(A|B) \tag{2.12}$$

From this it can be concluded that the condition for the independence of A and B can also be given either as $P(A|B) = P(A)$ or as $P(A \cap B) = P(A)P(B)$. Note that $P(A \cap B) = P(A)P(B|A)$ holds whether or not A and B are independent, but $P(A \cap B) = P(A)P(B)$ holds only when A and B are independent.

Now, events A and B are said to be independent if

$$P(A \cap B) = P(A)P(B) \tag{2.13}$$

2.5 Exclusive Events

An event is a well defined collection of some sample points in the sample space. Two events A and B in a universal sample space S are said to be exclusive events provided $A \cap B = \phi$. If A and B are exclusive events, then it is not possible for both events to occur on the same trail.

A list of events $A_1, A_2, \ldots, A_n$ is said to be mutually exclusive if and only if:

$$A_i \cap A_j = \begin{cases} A_i, & \text{if } i = j \\ \phi, & \text{otherwise} \end{cases} \tag{2.14}$$

So, a list of events is said to be composed of mutually exclusive events if no point in the sample space is common in more than one event in the list.

A list of events $A_1, A_2, \ldots, A_n$ is said to be collectively exhaustive, if and only if:

$$A_1 \cup A_2 \cup \ldots \cup A_n = S \tag{2.15}$$

2.6 Bayes' Rule

Bayes' rule is mainly helpful in determining a events probability from conditional probability of other events (exhaustive) in the sample space. Let $A_1, A_2, \ldots, A_k$ be mutually exclusive events whose union is the sample space S. Then for any event A,

$$P(A) = \sum_{i=1}^{n} P(A|A_i)P(A_i) \tag{2.16}$$

The above relation is also known as the theorem of total probability.

In some experiments, a situation often arises in which the event A is known to have occurred, but it is not known directly which of the mutually exclusive and collectively exhaustive events $A_1, A_2, \ldots, A_n$ has occurred. In this situation, to evaluate $P(A_i|A)$, the conditional probability that one of these events A_i occurs, given that A occurs. By applying the definition of conditional probability followed by the use of theorem of total probability, it comes out:

$$\begin{aligned} P(A_i|A) &= \frac{P(A_i|A)}{P(A)} \\ &= \frac{P(A|A_i)P(A_i)}{\sum_j P(A|A_j)P(A_j)} \end{aligned} \tag{2.17}$$

This relation is known as Bayes' rule. From these probabilities of the events $A_1, A_2, \ldots, A_k$ which can cause A to occur can be established. Bayes' theorem makes it possible to obtain $P(A|B)$ from $P(B|A)$, which in general is not possible.

2.7 Random Variables

When a real number is assigned to each point of a sample space, i.e. each sample point has a single real value. This is a function defined on the sample space. The result of an experiment which assumes these real-valued numbers over the sample space is called a random variable. Actually this variable is a function defined on the sample space.

A random variable defined on a discrete sample space is called a discrete random variable, and a stochastic variable defined on a continuous sample space and takes on a uncountable infinite number of values is called a continuous random variable.

In general a random variable is denoted by a capital letter (e.g. X, Y) whereas the possible values are denoted by lower case letter (e.g. x, y).

A random variable partitions its sample space into mutually exclusive and collectively exhaustive set of events. Thus, for a random variable X, and a real number x, let's define a A_x to be the subset of S consisting of all sample points s to which the random variable X assign the value x:

$$A_x = \{s \in S | X(s) = x\} \tag{2.18}$$

It is implied that $A_x \cap A_y = \phi$ if $x \neq y$, and that:

$$\bigcup_{x \in \Re} A_x = S \tag{2.19}$$

The collection of events A_x for all x defines an event space.

2.7.1 Discrete Random Variables

When the state space is discrete, and the random variable could take on values from a discrete set of numbers, the random variable is either finite or countable. Such random variables are known as discrete random variables. A random variable defined on a discrete sample space will be discrete, but it is possible to define a discrete random variable on a continuous sample space. For example, for a continuous sample space S, the random variable defined by $X(s) = 1$ for all $s \in S$ is discrete.

Let X be a random variable which take the values from sample space $\{x_1, x_2, \ldots, x_n\}$. If these values are assumed with probabilities given by

$$P\{X = x_k\} = f(x_k) \tag{2.20}$$

This is also known as frequency (or mass) function. In general, a function $f(x)$ is a mass function if

$$f(x) \geq 0 \tag{2.21}$$

and

$$\sum_x f(x) = 1 \tag{2.22}$$

where the sum is to be taken over all possible values of x.

The following properties hold for the probability mass function (pmf):

(a) As $P(X = x_k)$ is a probability, $0 \le P(X = x_k) \le 1$ must hold for all $s \in \Re$.
(b) Since the random variable assigns some value $x \in \Re$ to each sample point $x \in S$, the following must satisfy:

$$\sum_x f(x) = 1 \tag{2.23}$$

Let's move to compute the probability of the set $\{s|X(s) \in A\}$ for some subset A of $\Re$ other than a one-point set. It can be shown that:

$$\{s|X(s) \in A\} = \bigcup_{x_i \in A} \{s|X(s) = x_i\} \tag{2.24}$$

If $f(x)$ denotes the probability mass function of random variable X, then from above equation we have:

$$P(X \in A) = \sum_{x_i \in A} f(x_i) \tag{2.25}$$

The function $F(x)$, defined by:

$$\begin{aligned} F(x) &= P(-\infty < X \le x) \\ &= P(X \le x) \\ &= \sum_{x_i \le x} f(x_i) \end{aligned} \tag{2.26}$$

is called the probability distribution function or the cumulative distribution function (CDF) of the random variable X. It follows from this definition that:

$$\begin{aligned} P(a < X \le b) &= P(X \le b) - P(X \le a) \\ &= F(b) - F(a) \end{aligned} \tag{2.27}$$

Several properties of $F(x)$ follows directly from its definition:

1. $0 \le F(x) \le 1$ for $-\infty < x < \infty$, this follows from definition.
2. $F(x)$ is a monotone nondecreasing function of x. This follows by observing that the interval $(-\infty, x_1]$ is contained in the interval $(-\infty, x_2]$ whenever $x_1 \le x_2$ and hence:

$$F(x_1) \le F(x_2) \tag{2.28}$$

3. $\lim_{x\to\infty} F(x) = 1$ and $\lim_{x\to-\infty} F(x) = 0$
4. $F(x)$ has a positive jump equal to $f(x_i)$ at $i = 1, 2, \ldots$ and in the interval $[x_i, x_{i+1})$, $F(x)$ has constant value.

The distribution function is obtained from the density function by noting that

$$F(x) = P\{X \leq x\} = \sum_{u \leq x} f(u) \tag{2.29}$$

When X takes values from discrete sample space. The above equations are valid.

The cumulative distribution function contains most of the interesting information about the underlying probabilistic system, and this is used extensively. Often the concepts of sample space, event space, and probability measure, which are fundamental in building the theory of probability, will fade into the background, and functions such as the distribution function or the probability mass function become the most important entities.

2.7.1.1 Discrete Mathematical Distributions

Mostly used discrete probability distributions are as followed:

1. *Bernoulli distribution*
   The Bernoulli pmf is the density function of a discrete random variable X having 0 and 1 as its only possible values. It originates from the experiment consisting of a single trial with two possible outcomes. Mathematically, it is given by:

$$\begin{aligned} p_X(0) &= P(X = 0) = q \\ p_X(1) &= P(X = 1) = p \end{aligned} \tag{2.30}$$

   where $p + q = 1$. The corresponding CDF is given by:

$$F(X) = \begin{cases} 0, & \text{for } x < 0 \\ q, & \text{for } 0 \leq x < 1 \\ 1, & \text{for } x \geq 1 \end{cases} \tag{2.31}$$

2. *Binomial distribution*
   In a series of Bernoulli trails, the number of successes (or failures) out of total number of trials follows the Binomial distribution. Consider a sequence of n independent Bernoulli trials with probability of success equal to p on each trial. Let Y_n denote the number of successes in n trials. The domain of the random variable Y_n is all the n-tuples of $0's$ and $1's$, and the image is $\{0, 1, \ldots, n\}$. The value assigned to an n-tuple by Y_n simply corresponds to the number of $1's$ in the n-tuple.

$$p_k = P(Y_n = k) = \begin{cases} C(n,k)p^k q^{n-k}, & \text{for } 0 \le k \le n \\ 0, & \text{otherwise} \end{cases} \quad (2.32)$$

The above equation gives the probability of k 'successes' in n independent trials, where each trial has probability p of success.

3. *Geometric distribution*
Let's consider a sequence of Bornoulli trials, and count the number of trial until the first "success" occurs. Let 0 denote a failure and 1 denote a success, then the sample space of these trials consists of all binary strings with an arbitrary number of $0's$ followed by a single 1:

$$S = \{0^{i-1}1 | i = 1, 2, 3, \ldots\} \quad (2.33)$$

Note that this sample space has a countably infinite number of sample points. Let define a random variable Z on this sample space so that the value assigned to the sample points $0^{i-1}1$ is i. Thus Z is the number of trials up to and including the first success. Therefore, Z is a random variable with image $\{1, 2, 3, \ldots\}$, which is a countably infinite set. To find the pmf of Z, we note that the event $[Z = i]$ occurs if and only if we have a sequence of $i - 1$ failures followed by one success. This is a sequence of independent Bernoulli trails with probability of success equal to p. Hence, we have:

$$\begin{aligned} p_Z(i) &= q^{i-1}p \\ &= p(1-p)^{i-1}, \quad \text{for } i = 1, 2, 3, \ldots \end{aligned} \quad (2.34)$$

The geometric distribution has an important property, known as the Markov (or memoryless) property. This is the only discrete distribution with this property. To illustrate this property, consider a sequence of Bernoulli trials and let Z represent the number of trials until the first success. Now assume that we have observed a fixed number n of these trials and found them all to be failures. Let Y denote the number of additional trails that must be performed until the first success. Then $Y = Z - n$, and the conditional probability is:

$$\begin{aligned} q_i &= P(Y = i | Z > n) \\ &= P(Z - n = i | Z > n) \\ &= P(Z = n + i | Z > n) \\ &= \frac{P(Z = n + i \text{ and } Z > n)}{P(Z > n)} \\ &= \frac{P(Z = n + i)}{P(Z > n)} \\ &= \frac{pq^{n+i-1}}{1 - (1 - q^n)} \\ &= \frac{pq^{n+i-1}}{q^n} \\ &= pq^{i-1} \\ &= p_Z(i) \end{aligned} \quad (2.35)$$

We see that condition on $Z > n$, the number of trails remaining until the first success, $Y = Z - n$, has the same pmf as Z had originally.

4. *Negative binomial distribution*
   In geometric pmf, Bernoulli trials until the first success are observed. If r success need to be observed, then the process results in negative binomial pmf. Negative binomial pmf is given as:

$$
\begin{aligned}
&p_r(n) = p^r C(-r, n-r)(-1)^{n-r}(1-p)^{n-r} \\
&\text{where} : n = r, r+1, r+2, \ldots
\end{aligned}
\tag{2.36}
$$

   As negative binomial is generalization of Geometric pmf, for $r = 1$, this reduces to Geometric pmf.

5. *Poisson distribution*
   Let's observe the arrival jobs to a large computing center for the interval $(0, t]$. It is reasonable to assume that for each small interval of time Δt the probability of a new job arrival is $\lambda.\Delta t$, where λ is a constant that depends upon the user population of the computing center. If Δt is sufficiently small, then the probability of two or more jobs arriving in the interval of duration Δt may be neglected. We are interested in calculating the probability of k jobs arriving in the interval of duration t.
   Suppose that the interval $(0, t]$ is divided into n sub-intervals of length t/n, and suppose further that the arrival of a job in any given interval is independent of the arrival of a job in any other interval. Then for a sufficiently large n, we can think of the n intervals as constituting a sequence of Bernoulli trials with the probability of success $p = \lambda t/n$. It follows that the probability of k arrivals in a total of n intervals each with a duration t/n is approximately given by:

$$
b\left(k; n, \frac{\lambda t}{n}\right) = C(n,k)\left(\frac{\lambda t}{n}\right)^k \left(1 - \frac{\lambda t}{n}\right)^{n-k}, \quad k = 0, 1, \ldots, n \tag{2.37}
$$

   Since the assumption that the probability of more than one arrival per interval can be neglected is reasonable only if t/n is very small, we will take the limit of the above probability mass function as n approaches ∞.

$$
\begin{aligned}
b\left(k; n, \frac{\lambda t}{n}\right) &= \frac{n(n-1)(n-2)\cdots(n-k+1)}{k!n^k}(\lambda t)^k\left(1 - \frac{\lambda t}{n}\right)^{(n-k)} \\
&= \frac{n}{n}\cdot\frac{n-1}{n}\cdots\frac{n-k+1}{n}\cdot\frac{(\lambda t)^k}{k!}\cdot\left(1 - \frac{\lambda t}{n}\right)^{-k}\cdot\left(1 - \frac{\lambda t}{n}\right)^{n}
\end{aligned}
\tag{2.38}
$$

   We are interested in what happens to this expression as n increases, because then the subinterval width approaches zero, and the approximation involved gets better and better. In the limit as n approaches infinity, the first k factors approach unity, the next factor is fixed, the next approaches unity, and the last factor becomes:

$$\lim_{n\to\infty}\left(\left[1-\frac{\lambda t}{n}\right]^{-n/\lambda t}\right)^{-\lambda t} \tag{2.40}$$

Setting $-\lambda t/n = h$, this factor is:

$$\lim_{h\to 0}\left[(1+h)^{1/h}\right]^{-\lambda t} = e^{-\lambda t} \tag{2.41}$$

Since the limit the bracket is the common definition of e. Thus, the binomial probability mass function approaches:

$$\frac{e^{-\lambda t}(\lambda t)^k}{k!}, \quad k = 0, 1, 2, \ldots \tag{2.42}$$

Now replacing λt by a single parameter α, we get the well-known Poisson pmf:

$$f(k;\alpha) = e^{-\alpha}\frac{\alpha^k}{k!}, \quad k = 0, 1, 2, \ldots \tag{2.43}$$

It can be seen that Binomial probability mass function approaches Poisson probability mass function when n is large and p is small:

$$C(n,k) = p^k(1-p)^{n-k} \simeq e^{-\alpha}\frac{\alpha^k}{k!}, \quad \text{where } \alpha = np \tag{2.44}$$

6. *Hypergeometric distribution*
   In Binomial distribution, probability of occurance of events remain same during each experiment. In experiments such as drawing samples from a fixed set of samples, binomial corresponds to 'sampling with replacement'. But in some experiments, the chance of occurance of events changes with the course of experimentations. The Hypergeometric distribution is obtained while 'sampling without replacement'.
   Suppose we select a random sample of n components from a box containing N components, d of which are known to be defective. For the first component selected, the probability that it is defective is given by d/N, but for the second selection it remain same if the first is replaced. Otherwise, this probability is $(d-1)/(N-1)$ or $(d)/(N-1)$, depending on whether or not a defective component was selected in first experiment. In this experiment constant chances of occurance, as in Bernoulli trials, is not satisfied. The probability distribution of such kind of experiments are referred as Hypergeometric. Hypergeometric probability mass function, $h(k;n,d,N)$, defined to be the probability of choosing k defective components in a random sample of n

components, chosen without replacement, from a total of N components, d of which are defective. The sample space of the experiment consist of $C(N,n)$ sample points. The k defectives can be selected from d defectives in $C(d,k)$ ways, and $(n-k)$ non-defective components may be selected from $(N-d)$ non-defectives in $C(N-d,n-k)$ ways. The whole sample of n components with k defectives can be selected in $C(d,k)\cdot C(N-d,n-k)$ ways. Assuming an equiprobable sample space, the required probability is:

$$h(k;n,d,N)=\frac{C(d,k)C(N-d,n-k)}{C(N,n)},\quad \max(0,d+n-N)\leq k\leq \min(d,n) \tag{2.45}$$

7. *Uniform distribution*
Let X be a discrete random variable with a finite set of image $\{x_1,x_2,\ldots,x_N\}$. When all the image elements has equal chance of occurance, then probability mass function is given as:

$$p_X(x_i)=\begin{cases}\frac{1}{N}, & x_i \text{ in the image of } X\\ 0, & \text{otherwise}\end{cases} \tag{2.46}$$

Such a random variable is said to have a discrete uniform distribution. This distribution plays an important role in the theory of random numbers and its application to Monte-Carlo simulation. It may be noted that the concept of uniform distribution cannot be extended to a discrete random variable with a countably infinite image, $\{x_1,x_2,\ldots\}$. The requirements that $\sum p_X(x_i)=1$ and $p_X(x_i)=$ constant (for $i=1,2,\ldots$) are incompatible.
Let X take on the values $\{1,2,\ldots,N\}$ with $p_X(i)=1/N, 1\leq i\leq N$, then its distribution function is given by:

$$F(t)=\sum_{i=1}^{N}p_X(i) \tag{2.47}$$

2.7.2 Continuous Random Variables

In the previous section, we saw random variables and their distributions. In physical systems, such random variables denote the number of objects of certain type, such as number of failures detected during periodic inspection, or the number of call arrival at telephone exchange in a given time etc.

Many situations require the use of random variables that are continuous rather than discrete. As described earlier, a random variable is a real-valued function on the sample space S. When the sample space S is not countable, not every subset of the sample space is an event that can be assigned a probability.

A random variable X on a sample space is a function $X : S \rightarrow \Re$ that assigns a real number $X(s)$ to each sample point $s \in S$, such that for every real number x, the set $\{s|X(s) \le x\}$ is an event.

The distribution function F_X of a random variable X is defined to be the function

$$F_X(x) = P(X \le x), \quad -\infty < x < \infty \tag{2.48}$$

The subscript X is used to indicate the random variable under consideration. When there is no ambiguity the subscript will be dropped, and $F_X(x)$ will be denoted by $F(x)$.

The distribution function of a discrete random variable grows only by jumps as described in last section. The distribution function of a continuous random variable has no jumps, but grows continuously. Thus, a continuous random variable X is characterized by a distribution function $F_X(x)$ that is a continuous function of x for all x i.e. $-\infty < x < \infty$.

For continuous random variable, the random variable X takes any one particular value is in general zero. The probability that X is in between two different values is meaningful. In fact "$a < X \le b$" is the event corresponding to the set $(a, b]$.

For a continuous random variable, X, $f(x) = \mathrm{d}F(x)/\mathrm{d}x$ is called the probability density function (pdf) of X.

The pdf enables us to obtain the CDF by integrating under the pdf:

$$F_X(x) = P(X \le x) = \int_{-\infty}^{x} f_X(t)\mathrm{d}t, \quad -\infty < x < \infty \tag{2.49}$$

Other probabilities of interest are obtained as:

$$\begin{aligned} P(X \in (a, b]) &= P(a < X \le b) \\ &= P(X \le b) - P(X \le a) \\ &= \int_{-\infty}^{b} f_X(t)\mathrm{d}t - \int_{-\infty}^{a} f_X(t)\mathrm{d}t \\ &= \int_{a}^{b} f_X(t)\mathrm{d}t \end{aligned} \tag{2.50}$$

The pdf, $f(x)$, satisfies the following properties:

1. $f(x) \ge 0$ for all x.
2. $\int_{-\infty}^{\infty} f_X(x)\mathrm{d}x = 1$

It should be noted that, unlike the probability mass function, the values of the pdf are not probabilities and thus it is acceptable if $f(x) > 1$ at a points x.

2.7.2.1 Continuous Mathematical Distributions

Mostly used continuous probability distribution are as followed:

1. *Exponential distribution*
   The CDF of exponential distribution is given by:

$$F(x) = \begin{cases} 1 - e^{-\lambda x}, & \text{for } 0 \leq x < \infty \\ 0, & \text{otherwise} \end{cases} \quad (2.51)$$

   If a random variable X possesses CDF given by above equation, the pdf of X is given by:

$$f(x) = \begin{cases} \lambda e^{-\lambda x}, & \text{for } x > 0 \\ 0, & \text{otherwise} \end{cases} \quad (2.52)$$

   This distribution is also called negative exponential distribution. This distribution is widely used in applications such as reliability theory and queueing theory. Reasons for its wide use include its memoryless property (this result in analytical tractability) and its relation to the discrete Poisson and modified geometric distributions. The following random variables will often be modeled as exponential (provided experimental validation):

   (a) Time between two successive job arrivals to a computing center
   (b) Service time at a server in a queueing network
   (c) Time to failure of a electronic component
   (d) Time to repair a faulty component

2. *Hypoexponential distribution*
   Some processes can be divided into sequential phases for mathematical representation. If the time the process spends in each phase is independent and exponentially distributed, then the overall time is hypoexponentially distributed. The distribution has r parameters, one for each of its distinct phases.
3. *Erlang and Gamma distribution*
   When r sequential phases have independent identical exponential distributions, then the resulting density is known as r-stage Erlang. Mathematically, it's pdf is given as:

$$f(t) = \frac{\lambda^r t^{r-1} e^{-\lambda t}}{(r-1)!}, \quad t > 0, \quad \lambda > 0, \quad r = 1, 2, 3, \ldots \quad (2.53)$$

   and the CDF is given as:

$$F(t) = 1 - \sum_{k=0}^{r-1} \frac{(\lambda t)^k}{k!} e^{-\lambda t}, \quad t \geq 0, \quad \lambda > 0, \quad r = 1, 2, 3, \ldots \tag{2.54}$$

If r takes non integer values, then the process results in Gamma distribution. The density function is given as:

$$f(t) = \frac{\lambda^r t^{r-1} e^{-\lambda t}}{\Gamma r}, \quad t > 0, \quad \lambda > 0, \quad r > 0 \tag{2.55}$$

The Gamma function is defined as:

$$\Gamma n = \int_0^\infty x^{n-1} e^{-x} \mathrm{d}x, \quad n > 0 \tag{2.56}$$

And another useful identity is:

$$\int_0^\infty x^{n-1} e^{-\lambda x} \mathrm{d}x = \frac{\Gamma n}{\lambda^n} \tag{2.57}$$

4. *Weibull distribution*
   Weibull distribution is widely used for statistical curve fitting of lifetime data. The distribution has been used to describe fatigue failure, vacuum tube failure and ball bearing failure. The density function is given as:

$$f(x; \lambda, k) = \begin{cases} \frac{k}{\lambda} \left(\frac{x}{\lambda}\right)^{k-1} e^{-\left(\frac{x}{\lambda}\right)^k}, & x \geq 0 \\ 0, & x < 0 \end{cases} \tag{2.58}$$

   where k>0 is the shape parameter, and λ>0 is the scale parameter of the density function.
5. *Normal distribution*
   This distribution is extremely important in statistical applications because of the central limit theorem, which states that, under very general assumption, the mean of a sample of n mutually independent random variables (having distributions with finite mean and variance) is normally distributed in the limit $n \to \infty$. It has been observed that errors of measurement often possess this distribution.
   The normal density has the well-known bell-shaped curve and is given by:

$$f(x) = \frac{1}{\sigma\sqrt{2\pi}} e^{-\left(\frac{x-u}{\sqrt{2}\sigma}\right)^2} \tag{2.59}$$

where $-\infty < x, \mu < \infty$ and $\sigma > 0$. Here μ stands for mean and σ for standard deviation. As the above integral of above function does not have close form, distribution function $F(x)$ does not have close form. So for every pair of limits a and b, probabilities relating to normal distributions are usually obtained numerically or normal tables.
CDF of normal distribution with zero mean ($\mu = 0$) and unity standard deviation ($\sigma = 0$) is given as:

$$F_X(x) = \frac{1}{\sqrt{2\pi}} \int_{-\infty}^{x} e^{\frac{-t^2}{2}} dt \tag{2.60}$$

2.8 Transforms

A transform can provide a compact description of a distribution, and it is easy to compute widely used properties such as mean, variance and other moments.

2.8.1 Probability Generating Function

Probability generating function (PGF) is a mathematical tool that simplifies computations involving integer-valued, discrete random variables. For a given nonnegative integer-valued discrete random variable X with $P(X = k) = p_k$, PGF is defined as:

$$G_X(z) = \sum_{i=0}^{\infty} p_i z^i = p_0 + p_1 z + p_2 z^2 + \cdots + p_k z^k + \cdots \tag{2.61}$$

Equation 2.12 looks similar to $z-$ transform of X. $G_X(z)$ converges for any complex number z such that $|z| < 1$. For $z = 1$, it is easy to prove that,

$$G_X(1) = 1 = \sum_{i=0}^{\infty} p_i \tag{2.62}$$

In many problems PGF $G_X(z)$ will be known or derivable without the knowledge of pmf of X. It will be shown in later sections that interesting quantities such as mean and variance of X can be estimated from PGF itself. One reason for the usefulness of PGF is found in the following theorem, which has been quoted here without proof.

Theorem 2.1 *If two discrete random variables* X and Y*have same PGF's, then they must have the same distributions and pmf's.*

It means if a random variable has same PGF as that of another random variable with a known pmf, then this theorem assures that the pmf of the original random variable must be the same.

PGF of some widely used distributions is given below:

1. *Bernoulli random variable*

$$\begin{aligned} G_X(z) &= P(X=0)z^0 + P(X=0)z^1 \\ &= q + pz \\ &= 1 - p + pz \end{aligned} \tag{2.63}$$

2. *Binomial random variable*

$$\begin{aligned} G_X(z) &= \sum_{k=0}^{n} C(n,k)p^k(1-p)^{n-k}z^k \\ &= (pz + 1 - p)^n \end{aligned} \tag{2.64}$$

3. *Poisson random variable*

$$\begin{aligned} G_X(z) &= \sum_{k=0}^{\infty} \frac{\alpha^k}{k!} e^{-\alpha} z^k \\ &= e^{-\alpha} e^{\alpha z} \\ &= e^{\alpha(z-1)} \\ &= e^{-\alpha(1-z)} \end{aligned} \tag{2.65}$$

4. *Uniform random variable*

$$\begin{aligned} G_X(z) &= \sum_{k=1}^{N} \frac{1}{N} z^k \\ &= \frac{1}{N} \sum_{k=1}^{N} z^k \end{aligned} \tag{2.66}$$

2.8.2 Laplace Transform

As PGF simplifies computation of integer-valued, discrete random variable, Laplace transform simplifies computation of real-valued, continuous random variables. For a given nonnegative real-valued, integrable, continuous random variable X with pdf $p_X(x)$, Laplace transform is defined as:

$$L[p_X(x)] = L(s) = \int_0^{\infty} e^{-sx} p_X(x) \mathrm{d}x \tag{2.67}$$

The transform can give a number of interesting parameters. For $s = 0$, it reduces to:

$$L(0) = \int_0^\infty P_X(x)dx \tag{2.68}$$

Laplace distribution of widely used random distributions, are given below:

1. *Exponential distribution*

$$\begin{aligned} L(s) &= \int_0^\infty e^{-sx}\lambda e^{-\lambda x}\mathrm{d}x \\ &= \frac{\lambda}{\lambda + s} \end{aligned} \tag{2.69}$$

2. *Erlang distribution*

$$\begin{aligned} L(s) &= \int_0^\infty e^{-sx}\frac{\lambda e^{-\lambda x}(\lambda x)^{r-1}}{(r-1)!}\mathrm{d}x \\ &= \left(\frac{\lambda}{\lambda + s}\right)^r \end{aligned} \tag{2.70}$$

3. *Gamma distribution*

$$\begin{aligned} L(s) &= \int_0^\infty e^{-sx}\frac{\lambda e^{-\lambda x}(\lambda x)^{r-1}}{\Gamma r}\mathrm{d}x \\ &= \left(\frac{\lambda}{\lambda + s}\right)^r \end{aligned} \tag{2.71}$$

2.9 Expectations

The distribution function $F(x)$ or the density $f(x)$(or pmf for a discrete random variable) complete characterizes the behavior of a random variable X. Frequently, more concise description such as a single number or a few numbers, rather than an entire function. One such number is the *expectation* or the *mean*, denoted by $E[X]$. Similarly, others are *median*, *mode*, and *variance* etc. The mean, median and mode are often called *measures of central tendency* of a random variable X.

Definition 2.1 The expectation, $E[X]$, of a random variable X is defined by:

$$E[X] = \begin{cases} \sum_i x_i p(x_i), & \text{for discrete,} \\ \int_{-\infty}^{\infty} x f(x)\mathrm{d}x, & \text{for continuous,} \end{cases} \tag{2.72}$$

Equation 2.21 valid provided that the relevant sum or integral is absolutely convergent; that is,

$$\sum_i |x_i| p(x_i) < \infty$$
$$\int_{-\infty}^{\infty} |x| f(x) \mathrm{d}x < \infty \tag{2.73}$$

If these sum or integral are not absolutely convergent, then $E[X]$ does not exist.

Example 2.1 Let X be a continuous random variable with an exponential density given by:

$$f(x) = \lambda e^{-\lambda x}, \forall x > 0 \tag{2.74}$$

Expectation of X is evaluated as:

$$E[X] = \int_{-\infty}^{\infty} x f(x) \mathrm{d}x = \int_{0}^{\infty} \lambda x e^{-\lambda x} \mathrm{d}x \tag{2.75}$$

Let $z = \lambda x$, then $\mathrm{d}z = \lambda \mathrm{d}x$, putting these in above equation:

$$E[X] = \frac{1}{\lambda} \int_{0}^{\infty} z e^{-z} \mathrm{d}z \tag{2.76}$$

Please note definition of Gamma function:

$$\int_{0}^{\infty} x^n e^{-x} \mathrm{d}x = \Gamma(n+1), \forall n \in \Re \tag{2.77}$$

Using this $E[X]$ reduces to

$$E[X] = \frac{1}{\lambda} \Gamma 2 = \frac{1}{\lambda} 1! = \frac{1}{\lambda} \tag{2.78}$$

This gives a widely used result in reliability engineering. If a component obeys an exponential failure law with parameter λ, then its expected life, or its mean time to failure (MTTF), is $\frac{1}{\lambda}$.

2.10 Operations on Random Variables

While dealing with random variables, situation often arises which requires addition, maximum, minimum, mean, median, etc. In this section, these are discussed.

Let's determine the *pdf* of random variable Z, where $Z = X_1 + X_2 + \cdots + X_n$. $X_1, X_2, \ldots$ are independent random variables with known *pdf*. On a $n-$dimensional event space, this event is represented by all the events on the plane $X_1 + X_2 + \cdots + X_n = t$. The probability of this event may be computed by adding the probabilities of all the event points on this plane.

$$P(Z = t) = \sum_{\{x_1, x_2, \ldots, x_n\}} P(X_1 = x_1, X_2 = x_2, \ldots X_n = x_n; x_1 + x_2 \cdots + x_n = t)$$
$$p_z(t) = p_1(t) \otimes p_2(t) \otimes \cdots \otimes p_n(t) \tag{2.79}$$

This summation turn out to be convolution (discrete or continuous).

Let $Z_1, Z_2, \ldots, Z_n$ be random variables obtained by permuting the set, $X_1, X_2, \ldots, X_n$ so as to be in increasing order.

$$\begin{aligned} Z_1 &= \min\{X_1, X_2, \ldots, X_n\} \\ &\text{and} \\ Z_n &= \max\{X_1, X_2, \ldots, X_n\} \end{aligned} \tag{2.80}$$

The random variable Z_k is called $k-$th order statistic. To derive the distribution function of Z_k, note that the probability that exactly j of the X_i's lie in $(-\infty, z]$ and $(n-j)$ lie in (z, ∞) is:

$$C(n, j)F^j(z)\left[1 - F^{n-j}(z)\right] \tag{2.81}$$

since the binomial distribution with parameters n and $p = F(z)$ is applicable. Then:

$$\begin{aligned} F_{Z_k}(z) &= P(Z_k \leq z) \\ &= P(\text{``at least } k \text{ of the } X_i\text{'s lie in the interval } (-\infty, z)\text{''}) \\ &= \sum_{j=k}^{n} C(n, j)F^j(z)[1 - F^{n-j}(z)], \; -\infty < z < \infty \end{aligned} \tag{2.82}$$

In particular, the distribution function of Z_1 and Z_n can be obtained from 2.82 as:

$$\begin{aligned} F_{Z_1}(z) &= 1 - [1 - F(z)]^n \\ F_{Z_n}(z) &= F(z)^n \end{aligned} \tag{2.83}$$

2.11 Moments

Moment is generalization of expectation. In expectation, random variable along with *pdf* is summed (or integrated) over the entire sample space. Replacing random variable with another function of random variable, gives moment.

$$E[g(X)] = \begin{cases} \sum_i g(x_i)p_X(x_i), & \text{for discrete,} \\ \int_{-\infty}^{\infty} g(x)f_X(x)\mathrm{d}x, & \text{for continuous,} \end{cases} \tag{2.84}$$

Above equation is valid provided the sum or integral is absolutely convergent.

A interesting case of interest is the power function of X, i.e. $g(X) = X^k$. For $k = 1, 2, 3, \ldots$ moment is known as the kth moment of the random variable X. For $k = 1$, the first moment, $E[X]$, is the ordinary mean or expectation of X.

Isomorphic property: If two random variables X and Y have matching corresponding moments of all orders, then X and Y have the same distribution.

Sum and product of two random variables are of special interest. Hence expectation of these are discussed here.

Theorem 2.2 *Let X and Y be two random variables. Then the expectation of their sum is the sum of their expectations if* $E[Z] = E[X + Y] = E[X] + E[Y]$.

The above theorem does not require that X and Y be independent. It can be generalized to the case of n variable:

$$E\left[\sum_{i=1}^{n} X_i\right] = \sum_{i=1}^{n} E[X_i] \tag{2.85}$$

and

$$E\left[\sum_{i=1}^{n} a_i X_i\right] = \sum_{i=1}^{n} a_i E[X_i] \tag{2.86}$$

where $a_1, \ldots, a_n$ are constants.

Theorem 2.3 *Let X and Y be two independent random variables. Then the expectation of their product is the product of their expectations.* $E[XY] = E[X]E[Y]$.

2.12 Summary

Probability theory forms the basis of all probabilistic modeling. Dependability modeling is also based on these concepts. A complete treatment to this subject is beyond the scope of this book. In this chapter only basic concepts and theorems are refreshed. Interested readers may refer text dedicated to probability theory.

Chapter 3
Stochastic Processes and Models

3.1 Introduction

The theory of stochastic processes is generally defined as the "dynamic" part of the probability theory, in which one studies a collection of random variables (called a *stochastic process*) from the point of view of their interdependence and limiting behavior. One is observing a stochastic process whenever one examines a process developing in time in a manner controlled by probabilistic laws. Examples of stochastic processes are provided by the path of a particle in Brownian motion, the growth of a population such as a bacterial colony and the fluctuating number of particles emitted by a radioactive source, etc.

The measurements of an experiment, weather forecast, behavior of servo-mechanism, behavior of communication system, market price fluctuation and brain-wave records—all faces a problem of random variation to which theory of stochastic processes[1] may be relevant. The theory is an essential part of such diverse fields as statistical physics, theory of population growth, communication and control theory, operation research (management science), computer performance analysis and dependability analysis.

Definition 3.1 (*stochastic process*) A stochastic process is a family of random variables $\{X(t)|t \in T\}$, defined on a given probability space, indexed by the parameter *t*, where *t* varies over an index set *T*.

The values assumed by the random variable $X(t)$ are called states, and the set of all possible values forms the state space of the process.

If the state space of a stochastic process is discrete, then it is called a discrete-state process, often referred to as a chain. In this case, the state space is often assumed to be integers. If the state space is continuous, then we have a

[1] The word "stochastic" is of Greek origin. In seventeenth century English, the word "stochastic" had the meaning "to conjecture, to aim at mark". It is not quite clear how it acquired the meaning it has today of "pertaining to chance".

A. K. Verma et al., *Dependability of Networked Computer-based Systems*,
Springer Series in Reliability Engineering, DOI: 10.1007/978-0-85729-318-3_3,

continuous-state process. Similarly, if the index set T is discrete, then we have a discrete-parameter process; otherwise a continuous parameter process.

3.2 Classification of Stochastic Processes

For a given time $t = t_0$, process $X(t_0)$ is a simple random variable that describes the state of the process at time t_0. For a fixed number x_1, the probability of the event $[X(t_0) \leq x_1]$ gives the cumulative distribution function (CDF) of the random variable $X(t_0)$. Mathematically, this is given as:

$$F_{X(t_1)}(x_1) = P[X(t_1) \leq x_1] \tag{3.1}$$

$F(x_1; t_1)$ is known as the first-order distribution of the process $X(t)$. Given two time instants t_1 and $t_2, X(t_1)$ and $X(t_2)$ are two random variables on the same probability space. Their joint distribution is known as the second-order distribution of the process and is given by:

$$F(x_1, x_2; t_1, t_2) = P(X(t_1) \leq x_1, X(t_2) \leq x_2) \tag{3.2}$$

In general, the nth order joint distribution of the stochastic process $\{X(t), t \in T\}$ by:

$$F(\mathrm{x} : \mathrm{t}) = P[X(t_1) \leq x_1, \ldots, X(t_n) \leq x_n] \tag{3.3}$$

for all $x = (x_1, \ldots, x_n) \in \Re^n$ and $t = (t_1, \ldots, t_n) \in T^n$ such that $t_1 < t_2 < \cdots < t_n$. Many processes of practical interest, however, permit a much simpler description.

The processes can be classified based on time-shift, independence and memory, as follows:

1. A stochastic process $\{X(t)\}$ is said to be stationary in the strict sense if for $n \geq 1$, its nth-order joint CDF satisfies the condition:

$$F(x : t) = F(x : t + \tau) \tag{3.4}$$

   for all vector $\mathrm{x} \in \Re^{\mathrm{n}}$ and $t \in T^n$, and all scalars τ such that $t_i + \tau \in T$. The notation $t + \tau$ implies that the scaler τ is added to all components of vector t. Let $\mu(t) = E[X(t)]$ denote the time-dependent mean of the stochastic process. $\mu(t)$ is often called the ensemble average of the stochastic process. Applying the definition of the strictly stationary process to the first-order CDF, $F(x; t) = F(x; t + \tau)$ for all τ. It follows that a strict-sense stationary process has a time-independent mean; that is, $\mu(t) = \mu$ for all $t \in T$.
2. A stochastic process $\{X(t)\}$ is said to be an independent process provided its nth-order joint distribution satisfies the condition:

$$F(x : t) = \prod_{i=1}^{n} F(x_i\colon t_i) = \prod_{i=1}^{n} P[X(t_i) \leq x_i] \tag{3.5}$$

3. A renewal process is defined to be a discrete-parameter independent process $\{X_n | n = 1, 2, \ldots\}$ where $X_1, X_2, \ldots,$ are independent, identically distributed, nonnegative random variables.

Consider a system in which the repair after failure is performed, requiring negligible time. The time between successive failures might be independent, identically distributed random variables $\{X_n | n = 1, 2, \ldots\}$ of a renewal process.

3.3 The Random Walk

Random walk has its origin in study of movement of a particle in fluid. But the random walk has been used in widely variety of applications such as modeling of insurance risk, escape of comets from the solar system, content of dam and queueing system, etc. Consider a particle which can move only in one dimension i.e. x-axis. At time $n = 1$ the particle undergoes a step or jump Z_1, where Z_1 is a random variable having a given distribution. At time $n = 2$ the particle undergoes a jump Z_2, where Z_2 is independent of Z_1 and with the same distribution, and so on. As the particle moves along a straight line and after one jump is at the position $X_0 + Z_1$, after two jumps at $X_0 + Z_1 + Z_2$ and, in general, after n jumps the position of the particle is given as $X_n = X_0 + Z_1 + Z_2 + \cdots + Z_n$. Here Z_i is a sequence of mutually independent, i.i.d. random variables. This can be represented as $X_n = X_{n-1} + Z_n$ for $n = 1, 2, \ldots$

Now for a particular case where the steps Z_i can only take the values $1, 0, -1$ with the probability:

$$\begin{aligned} P(Z_i = 1) &= p \\ P(Z_i = -1) &= q \\ P(Z_i = 0) &= 1 - p - q \end{aligned} \tag{3.6}$$

The above particular process is a stochastic process in discrete time with discrete state space. If the particle continues to move indefinitely according to above relation the random walk is said to be unrestricted. The motion of the particles may be restricted by use of barriers. These barriers could be absorbing or reflection barriers. Till now, we have restricted the discussion to one-dimensional jumps/ steps only. When the jumps are in two or three dimension, it results in two or three-dimensional random walk, respectively.

Example The escape of comets from the solar system. This example has been taken from Cox [1]. This problem was originally studied by Kendell. He has made an interesting application of the random walk to the theory of comets. Comets revolve around earth and during one revolution round the earth the energy of a comet undergoes a change brought about by the disposition of the planets. In successive revolutions the change in energy of the comet are assumed to be independent and identically distributed random variables $Z_1, Z_2, \ldots$. If initially the comet has positive energy X_0 then after n revolutions the energy will be

$$X_n = X_0 + Z_1 + Z_2 + \cdots + Z_n \tag{3.7}$$

If at any stage the energy X_n becomes zero or negative, the comet escapes from the solar system. Thus the energy level of the comet undergoes a random walk starting at $X_0 > 0$ with an absorbing barrier at 0. Absorption corresponds to escape from the solar system.

3.4 Markov Chain

Definitions 3.2 (*Markov Chain*) A discrete-state stochastic process $\{X(t), t \geq 0\}$ is called a Markov chain if, for any $t_1 < t_2 < \cdots < t_n$, the conditional probability of being in any state j is such that: [1–3]

$$\Pr\{X(t_n) = j|X(t_{n-1}) = i_{n-1}, \ldots, X(t_0) = i_0\} = \Pr\{X(t_n) = j|X(t_{n-1}) = i_{n-1}\} \tag{3.8}$$

This condition is called *Markov property*, which means the state of a Markov chain after a transition probabilistically depends only on the state immediately before it. In other words, at the time of a transition the entire past history is summarized by the current state (and implicitly by the current time t).

A homogeneous discrete-time Markov chain may be represented by one-step transition probability matrix P with elements:

$$p_{ij} = P\{X_{n+1} = x_j|X_n = x_i\} \tag{3.9}$$

where x_i represents the state of the system at discrete-time-step $t \in N$.

p_{ij} is the probability of x_j being the next state given that x_i is the current state. So all the entries of P satisfies:

$$\begin{aligned} &(1) \quad 0 < p_{ij} < 1 \quad \forall i,j \\ &(2) \quad \sum_j p_{ij} = 1 \quad \forall i \end{aligned} \tag{3.10}$$

Definition 3.3 (*Irreducible Markov Chain*) A Markov chain is said to be irreducible if every state can be reached from every other state in a finite number of steps [2]. In other words, for all $i, j \in I$, there is an integer $n > 1$ such that $p_{ij}(n) > 0$.

Definition 3.4 (*Recurrent State*) A state i is said to be recurrent if and only if, starting from state i, the process eventually returns to state i with probability one [2].

Definition 3.5 (*Mean recurrence time*) The mean recurrence time of recurrent state x_j is [4]

$$M_j = \sum m f_j(m) \tag{3.11}$$

where: $f_j(m)$ denote the probability of leaving state x_j and first returning to that same state in m steps.

If $M_j = \infty$ state x_j is recurrent null; otherwise $M_j < \infty$ and x_j is recurrent non-null.

Theorem *The states of an irreducible Markov chain are either all transient or all recurrent non-null or all recurrent null. If the states are periodic, then they all have the same period.*

Limiting probability distribution: The limiting probability distribution $\{p_j\}$ of a discrete-time Markov chain is given by:

$$p_j = \lim_{m \to \infty} p_j(m) \tag{3.12}$$

Theorem *In an irreducible and aperiodic homogeneous Markov chain, the limiting probabilities $\{p_j\}$ always exists and are independent of initial probability distribution.*

Also one of the following conditions hold:

- Every state x_j is transient or every state x_j is recurrent null, in which case $p_j = 0$ for all x_j and there exists no stationary distribution (even though the limiting probability distribution exists). In this case, the state space must be infinite.
- Every state x_j is recurrent non-null with $p_j > 0$ for all x_j, in which case the set $\{p_j\}$ is a limiting and stationary probability distribution and

$$p_j = \frac{1}{M_j} \tag{3.13}$$

In this case, the p_j are uniquely determined form the set of equations:

$$p_j = \sum_i p_i p_{ij} \quad \text{subject to} \sum_i p_i = 1 \tag{3.14}$$

If $p = (p_1, p_2, \ldots)$ is a vector of limiting probabilities, then they can be evaluated by:

$$p = pP \tag{3.15}$$

where P is the transition probability matrix. The vector P is called the steady-state solution of the Markov chain.

3.4.1 Markov processes with Discrete state in discrete time

In discrete time Markov chain, the process can change its state only at discrete time points. The time process spends in a given state will be investigated here. Due to Markov property, the next transition does not depend up on how this state is reached and how much time has passed in this state. Let process has already spent

n_0 time quanta in a given state. The random variable 'transition time' is denoted by X and the random variable 'transition time after n_0' as Y and $Y = X - n_0$. Let conditional probability of $Y = n_0$, given that $X > n_0$, be denoted by $Z(n)$.

$$\begin{aligned} Z(n) &= P(Y = n|X > n_0) \\ &= P(X - n_0 = n|X > n_0) \\ &= P(X = n + n_0|X > n_0) \\ &= \frac{P(X = n + n_0 \text{ and } X > n_0)}{P(X > n_0)} \\ &= \frac{P(X = n + n_0)}{P(X > n_0)} \end{aligned} \tag{3.16}$$

So, in a discrete time Markov process, the distribution of resident time in a state posses a unique property. That is, given that it has already spent a specified time n_0, does not affect distribution of residual time. Means, process does not have time memory.

For a discrete-time Markov chain, the only sojourn time distribution, which satisfies the memory less, sojourn time condition is the geometric distribution.

If the conditional probability defined above is invariant with respect to the time origin t_n, then the Markov chain is said to be homogeneous, i.e., for any t and t_n,

$$\Pr\{X(t) = j|X(t_n) = i_n\} = \Pr\{X(t - t_n) = j|X(0) = i_n\} \tag{3.17}$$

If the states in a Markov can change only at discrete time points, the Markov chain is called a *discrete-time Markov chain* (DTMC). If the transitions between states may take place at any instance, the Markov chain is called a *continuous-time Markov chain* (CTMC).

3.4.2 Markov Processes with Discrete States in Continuous Time

In the last section we saw Markov process with discrete states and defined at discrete time instant. Let the state space of the process remain discrete and parameter space $t = [0, \infty)$. As per the definition of Markov process, a discrete-state continuous-parameter (time) stochastic process $\{X(t), t \geq 0\}$ is a Markov process if

$$P[X(t) = x|X(t_n) = x_n, X(t_{n-1}) = x_{n-1}, \ldots, X(t_0) = x_0,] = P[X(t) = x|X(t_n) = x_n] \tag{3.18}$$

where

$$t_0 < t_1 < t_2, \ldots, < t_n < t$$
$$x_i \in \{\text{state-space of process } I\}$$

A homogeneous continuous-time Markov chain may be represented by set of states and an infinitesimal generator matrix Q where $Q_{ij}, i \neq j$ represents the exponentially distributed transition rate between x_i and x_j. The parameter of the exponential distribution of the sojourn time in state x_i is given by—Q_{ij}, where $Q_{ij} = -\sum_{i \neq j} Q_{ij}$.

Note that the entries of Q must satisfy:

$$\sum_j Q_{ij} = 0 \quad \forall i \tag{3.19}$$

Theorem *In a finite, irreducible, homogeneous continuous-time Markov chain, the limiting probabilities* $\{p_j\}$ *always exit and are independent of the initial probability distribution.*

The steady state probability vector $p = (p_1, p_2, \ldots)$ can be determined from the following equations:

$$\begin{aligned} &(1) \quad pQ = 0 \\ &(2) \quad \sum_i p_i = 1 \end{aligned} \tag{3.20}$$

For a finite state-space process, the sum of all state probabilities equals to unity. The evolution of Markov process over time can be realized using Chapman–Kolmogorov (C–K) equation. C–K enables to build up conditional *pdf* over 'long' time interval from those over the 'short' time intervals. The transition probabilities of a Markov chain $\{X(t), t \geq 0\}$ satisfy the C–K equation for all $i, j \in I$,

$$\begin{aligned} &p_{ij}(v,t) = \sum_{k \in I} p_{ik}(v,u) p_{kj}(u,t) \\ &\text{for} \\ &0 \leq v < u < t \end{aligned} \tag{3.21}$$

A Markov process transits from one state to other, this state transition is captured by state-transition matrix. Unlike in discrete time Markov process, in continuous time process state transition may occur at any time. These two condition imposes restriction on the probability distribution a Markov process may have. Considering the Markov property and these two conditions together with time of transition, it is clear that time the process spends in a given state before transition does not depends on the time it has already spent in that state.

Let process has already spent time t_0 in a given state. The random variable 'transition time' is denoted by X and the random variable 'transition time after t_0' as Y and $Y = X - t_0$. Let conditional probability of $Y \leq t$, given that $X > t_0$, be denoted by $Z(t)$.

$$\begin{aligned} Z(t) &= P(Y \le t | X > t_0) \\ &= P(X - t_0 \le t | X > t_0) \\ &= P(X \le t + t_0 | X > t_0) \\ &= \frac{P(X \le t + t_0 \text{ and } X > t_0)}{P(X > t_0)} \\ &= \frac{P(t_0 \le X \le t + t_0)}{P(X > t_0)} \end{aligned} \tag{3.22}$$

Let *pdf* of X is given as $f(x)$, then $Z(t)$ is given as

$$Z(t) = \frac{\int_{t_0}^{t+t_0} f(x)dx}{\int_t^{\infty} f(x)dx} \tag{3.23}$$

For homogeneous time process, one such distribution is exponential distribution which fulfill this restriction. This implies for homogeneous continuous time parameter Markov process time to transition follows exponential distribution.

The distinction between homogeneous CTMC and non-homogeneous CTMC is that the sojourn time distribution in the homogeneous CTMC is exponential distribution. The sojourn time distribution in the non-homogeneous case is quite complex.

3.5 Non-Markovian Processes

We have seen in the last section, in a Markov process future evolution depends only on the present state, i.e. it does not depend on how that state is reached and how much time has already elapsed in that state. Any process not fulfilling these properties are termed as non-Markovian process. Some of the non-Markov processes with some unique property are of special interest, some of them are discussed in this section.

3.5.1 Markov Renewal Sequence

Markov renewal sequences [3, 5] play an important role in the formulation of semi-Markov, Markov renewal and Markov regenerative process.

Definition 3.6 Let S be a discrete state space. A sequence of bivariate random variables $\{(Y_n, T_n), n \ge 0\}$ is called a Markov renewal sequence if:

1. $T_0 = 0, T_{n+1} \ge T_n T_n \in \Re^+; Y_n \in S \forall n \ge 0$
2. $\forall n \ge 0$,

$$\begin{aligned}
&\Pr\{Y_{n+1} = j, T_{n+1} - T_n \le x | Y_n = i, T_n, Y_{n-1}, T_{n-1}, \ldots, Y_0, T_0\} \\
&\quad = \Pr\{Y_{n+1} = j, T_{n+1} - T_n \le x | Y_n = i\} \\
&\quad = \Pr\{Y_1 = j, T_1 \le x | Y_0 = i\} \\
&\quad = k_{i,j}(x)
\end{aligned} \tag{3.24}$$

The first equation indicates the Markov property at time point $\{T_n\}$, and the second indicates the time homogeneity. The matrix $K(x) = \|k_{i,j}(x)\|$ is referred to as the *kernel* of the Markov renewal sequence. The definition of Markov regenerative process (MRGP) is based on Markov renewal sequences.

3.5.2 Markov Regenerative Processes

Markov regenerative processes are the processes with embedded Markov renewal sequences. The formal definition is as follows.

Definition 3.7 A stochastic process $\{Z(t), t \ge 0\}$ is called a Markov regenerative process if there exists a Markov renewal sequence $\{(Y_n, T_n), n \ge 0\}$ of random variables such that all the conditional finite distributions of $\{Z(T_n + t), t \ge 0\}$ given $\{Z(u), 0 \le u \le T_n, Y_n = i\}$ are the same as those of $\{Z(t), t \ge 0\}$ given $Y_0 = i$. [1, 3, 5–7]

The above definition implies that:

$$\Pr\{Z(T_n + t) = j | Z(u), 0 \le u \le T_n, Y_n = i\} = \Pr\{Z(t) = j | Y_0 = i\} \tag{3.25}$$

It also implies that the future of the Markov regenerative process $\{Z(t), t \ge 0\}$ from $\{t = T_n\}$ onwards depends on the past $\{Z(t), 0 \le t \le T_n\}$ only through Y_n.

Let $v_{i,j}(t) = \Pr\{Z(t) = j | Y_0 = i\}$. The matrix $V(t) = \|v_{i,j}(t)\|$ is referred to as conditional transient probability matrix of MRGPs. The following theorem gives the generalized Markov renewal equation satisfied by $V(t)$. For the sake of conciseness, $(K * V)(t)$ to denote the matrix whose element (i,j) is defined as follows:

$$(k * v)_{i,j}(t) = \sum_{h \in S} \int_0^t dk_{i,h}(u) v_{h,j}(t - u) \tag{3.26}$$

Means, $(K * V)(t)$ is a matrix of functions of t whose generic element is obtained as the row by column convolution of the matrix $K(t)$ and matrix $V(t)$.

Theorem 3.1 *Let* $\{Z(t), t \ge 0\}$ *be an MRGP with embedded Markov renewal sequence* $\{(Y_n, T_n), n \ge 0\}$ *with kernel* $K(\cdot)$ [1, 2, 3, 5, 6, 7]. *Let*

$$e_{i,j}(t) = \Pr\{Z(t) = j, T_1 > t | Y_0 = i\} \tag{3.27}$$

and $E(t) = \|e_{i,j}(t)\|$. *Then* $V(\cdot)$ *satisfies the following Markov renewal equation:*

$$V(t) = E(t) + \int_0^t V(t - x) dK(x) = E(t) + K(t) * V(t) \tag{3.28}$$

The proof of this theorem is shown in [6, 7]. Note that $E(t)$ contains information about the behavior of MRGP over the first "cycle" $(0, S_1)$. Thus this theorem relates the behavior of the process at time t to its behavior over the first cycle.

Consider an M/G/1 queuing system, let $Z(t)$ be the number of customers at time t, we can define the embedded Markov renewal sequence, $\{(Y_n, S_n), n \geq 0\}$, as $S_0 = 0$ and S_n is the time of the nth customer departure; and $Y_n = Z(S_n+)$ [6, 7]. Note that $\{Z(t), t \geq 0\}$ satisfies the property of Definition 3.7 Hence, it is Markov regenerative process.

3.6 Higher Level Modeling Formalisms

Model based dependability/performance evaluation of engineering systems is a powerful and inexpensive way of predicting the dependability/performance before the actual implementation. Its importance increases with the system complexity and criticality of applications. Markov models provide suitable framework for dependability, performance and performability [8, 9] evaluation. But, there are some difficulties in using Markov models [10].

1. State space grows much faster than the number of components in the system being modeled. A large state space can make a model difficult to specify correctly.
2. A Markov model of a system is sometimes far removed from the shape and general feel of the system being modeled. System designers may have difficulty in directly translating their problem into a Markov model.

These difficulties can be overcome by using a modeling technique that is more concise in its specification and whose form is close to a designer's intuition about what a model should look like. One of the most popular approach is to use Petri net based stochastic models. Molloy [11] showed stochastic Petri nets can be used to automatically generate an underlying Markov model, which can then be analyzed to yield results of interests. In this case the "user-level representation" of a system is translated into a different "analytical representation". The analytical representation is processed to evaluate results.

The analytical tractability of Markov models is based on the exponential assumption of the distribution of the holding time in a given state. This implies that the future evolution of the system depends only on the current state and, based on this assumption, simple and tractable equations can be derived for both transient and steady state analysis.

Nevertheless, the exponential assumption has been regarded as one of the main restrictions in the application of Markov models. In practice there is a very wide range of circumstances in which it is necessary to model phenomenon whose times to occurrence is not exponentially distributed. The hypothesis of exponential distribution thus allows the definition of models which can give a more qualitative rather than quantitative analysis of real systems. The existence of deterministic or other non-exponentially distributed event times, such as timer expiration, propagation delay, transmission of fixed length packets, etc. gives rise to stochastic models that are non-Markovian in nature.

In recent years considerable effort has been devoted to enrich Petri nets formalism in order to improve their capability to easily capture system behavior and to deal with generally distributed delays.

3.6.1 Petri Nets

Petri net is a directed, bipartite graph consisting of two kinds of nodes, called *places* and *transitions*, where arcs are either from a place to transition or from a transition to place [12–14]. Mathematically, Petri net structure is defined as a 5-tuple. Petri net: $N = \{P, T, I, O, M_0\}$, where:

- P is a finite set of "places"
- T is a finite set of "transitions"
- $I = (P \times T)$ defines the input function
- $O = (T \times P)$ defines the output function
- M_0 is the initial "marking" of the net, where a "marking" is the number of "tokens" contained in each place.

A transition t_i is said to be "enabled" by a marking m if and only if $I(t_i)$ is contained in m. Any transition t_i enabled by marking m_j can "fire". When it does, token(s) is removed from each place $I(t_i)$ and added to each place $O(t_i)$. This may result in a new marking m_k. If a marking enables more than one transition, the enabled transitions are said to be in conflict. Any of the enabled transition may fire first. This firing may disable transitions which were previously enabled.

Petri nets are generally represented graphically. Places are drawn as circles and transitions as bars. The input and output functions are represented by directed arcs from places to transitions and transitions to places, respectively. Tokens are represented by black dots or number inside places.

An example of Petri net model is given in Fig. 3.1. The PN consists of five places $\{p_1, p_2, p_3, p_4, p_5\}$ and 4 transitions $\{t_1, t_2, t_3, t_4\}$. The initial marking $M_0 =$ (1,0,0,1,0). In the present marking transition t_2 is only enabled. Firing of transition removes the token from place p_1 and deposit one token each in places p_2 and p_3. One of the possible firing sequence is $t_2, t_1, t_3, t_4, \ldots$

The analysis of Petri nets revolves around investigating the possible markings. The Petri net semantic does not state which of multiple simultaneously enabled transitions fires first, so a Petri net analysis must examine every possible firing order.

Petri nets can be used to capture the behavior of many real-world situations including sequencing, synchronization, concurrency, and conflict. The main feature which distinguishes PNs from queuing networks is the ability of the former to represent concurrent execution of activities. If two transitions are simultaneously

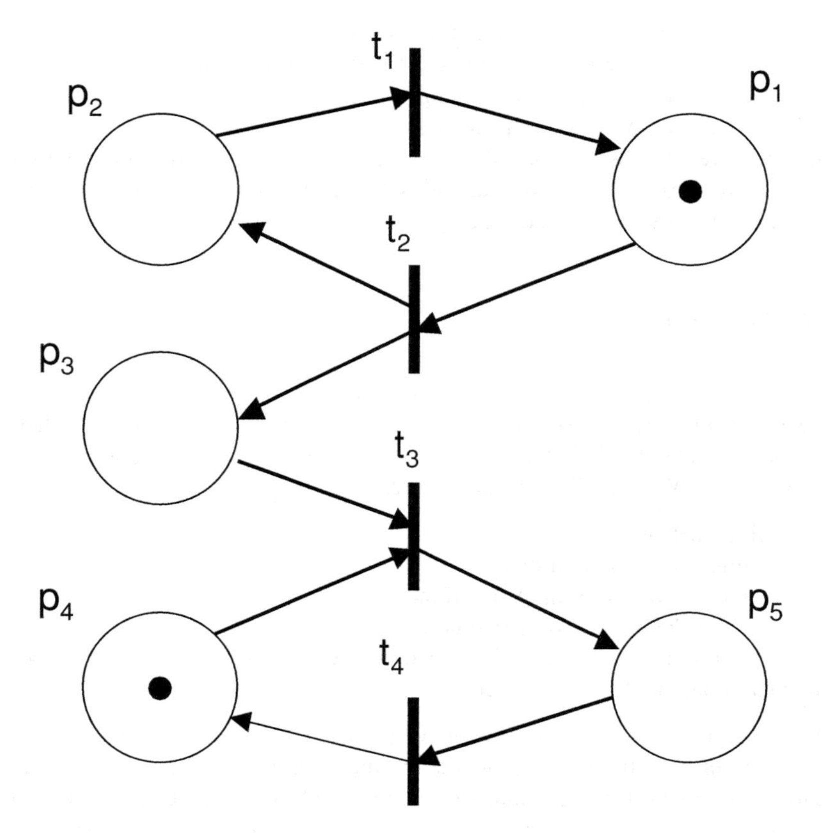

Fig. 3.1 Petri net

enabled, this means that the activities they represent are proceeding in parallel. Transition enabling corresponds to the *starting* of an activity, while transition firing corresponds to the *completion* of an activity. When the firing of a transition causes a previously enabled transition to become disabled, it means the interrupted activity was aborted before being completed.

3.6.2 Structural Extensions

Many extensions to PNs have been proposed to increase either the class of problems that can be represented or their capability to deal with the common behavior of real systems [10, 15]. These extension are aimed to increase, (1) modeling power, (2) modeling convenience, and (3) decision power [10]. Modeling power is the ability of a formalism to capture the details of a system. Modeling convenience is the practical ability to represent common behavior. Decision power is defined to be the set of properties that can be analyzed. The generally accepted conclusion is that increasing the modeling power decreases the decision power. Thus each possible extension to the basic PN formalism requires an in depth evaluation of its effect upon modeling and decision power.

Extensions which affect only modeling convenience can be removed by transforming an extended PN into an equivalent PN so they can usually be adopted without introducing any analytical complexity. These kind of extensions provide a powerful way to improve the ability of PNs to model real problems. Some extension of this type have proved so effective that they are now considered part of the standard PN definition. They are [10, 15]:

- arc multiplicity
- inhibitor arcs
- transition priorities
- marking-dependent arc multiplicity

Arc multiplicity is a convenient extension for representing a case when more than one token is to be moved to or from a place. The standard notation is to denote multiple arcs as a single arc with a number next to it giving its multiplicity.

Inhibitor arcs are another useful extension of standard PN formalism. An inhibitor arc from place p to transition t disables t for any marking where p is not empty. Graphically, inhibitor arcs connect a place to a transition and are drawn with a small circle instead of an arrowhead. It is possible to use the arc multiplicity extension in addition to inhibitor arcs. In this case a transition t is disable whenever place p contains at least as many tokens as the multiplicity of the inhibitor arc. Inhibitor arcs are used to model contention of limited resources to represent situations in which one activity must have precedence over another.

Another way to represent the latter situation is by using *transition priorities*, an extension in which an integer "priority level" is assigned to each transition. A transition is enabled only if no higher priority transition is enabled. However, the convenience of priorities comes at a price. If this extension is introduced, the standard PNs ability to capture the entire system behavior graphically is partially lost.

Practical situations often arise where the number of tokens to be transferred (or to enable a transition) depends upon the system state. These situations can be easily managed adopting *marking-dependent arc multiplicity*, which allows the multiplicity of an arc to vary according to the marking of the net. Marking dependent arc multiplicities allow simpler and more compact PNs than would be otherwise possible in many situations. When exhaustive state space exploration techniques are employed, their use can dramatically reduce the state space.

3.6.3 Stochastic Petri Nets

PNs lack the "concept of time" and "probability". Modeling power can be increased by associating random firing times with either the places or the transitions. When waiting times are associated with places, a token arriving into a place enables a transition only after the place's waiting time has elapsed. When waiting times are associated with transitions, an enabled transition fires only after the waiting time has elapsed, which is also referred as firing time.

Stochastic Petri net (SPN) models increase modeling power by associating exponentially distributed random firing times with the transitions [11]. A transition's firing time represents the amount of time required by the activity associated with the transition. It is counted from the instant the transition is enabled to the instant it actually fires, assuming that no other transition firing affects it.

An SPN example from Molloy [11] is taken. The SPN model is shown in Fig. 3.2. To illustrate derivation of CTMC (continuous time Markov chain) from this SPN, first all possible markings are estimated.

All possible markings of the SPN of Fig. 3.2 is shown in Table 3.1

Each marking corresponds to a state in the Markov chain. Possible transitions from each states corresponds to transitions of Markov chain. The equivalent Markov chain is shown in Fig. 3.3.

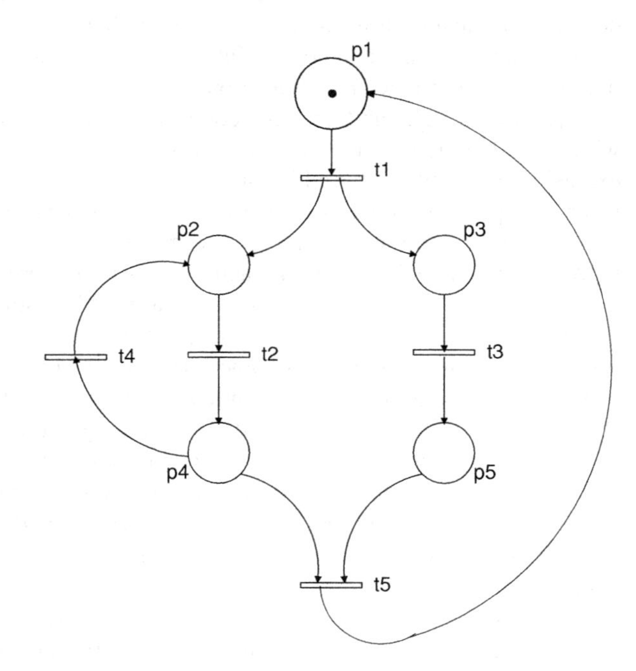

Fig. 3.2 Stochastic Petri net [11]

Table 3.1 Reachability graph of SPN [11]

	P1	P2	P3	P4	P5
M1	1	0	0	0	0
M2	0	1	1	0	0
M3	0	0	1	1	0
M4	0	1	0	0	1
M5	0	0	0	1	1

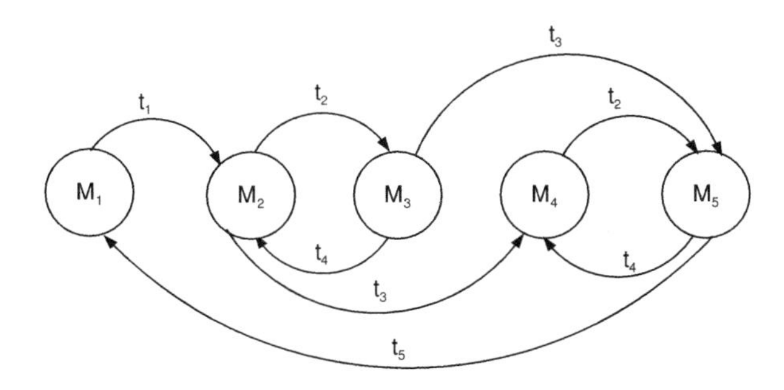

Fig. 3.3 Equivalent CTMC of the SPN [11]

Like basic PN models, SPN models can have more than one transition enabled at a time. To specify which transition will fire among all of those enabled in a marking, an "execution policy" has to be specified. Two alternatives are the "race policy" and the "pre-selection policy". Under the race policy, the transition whose timing time elapses first is assumed to be the one that will fire. Under the pre-selection policy, the next transition to fire in a given marking is chosen from among the enabled transitions using a probability distribution independent of their firing times. SPN models use the race policy.

3.6.4 Generalized Stochastic Petri Nets

Generalized stochastic petri nets (GSPNs), proposed by Marsan et al. [16], are an extension of Stochastic Petri nets obtained by allowing the transitions of the underlying PN to be immediate as well as timed. Immediate transitions (drawn as thin black bars) are assumed to fire in zero time once enabled. Timed transitions (represented by rectangular boxes or thick bars) are associated firing time just as in SPNs.

When both immediate and timed transitions are enabled in a marking, only the immediate transitions can fire; the timed transitions behave as if they were not enabled. When a marking m enables more than one immediate transition, it is necessary to specify a probability mass function according to which the selection of the first transition to fire is made. The markings of a GSPN can be classified into "vanishing" markings in which at least one immediate transition is enabled, and "tangible" markings, in which no immediate transitions are enabled. The reachability graph of a GSPN can be converted into a CTMC by eliminating vanishing markings and solved using known methods.

3.6.5 Stochastic Reward Nets (SRN)

Stochastic Reward Nets (SRN) introduce a stochastic extension into SPNs consisting of the possibility to associate reward rates with the markings. The reward rate definitions are specified at the net level as a function of net primitives like the number of tokens in a place or the rate of a transition. The underlying *Markov model* is then transformed into a *Markov reward model* thus permitting evaluation of not only performance and availability but also a combination of the two.

A stochastic reward net (SRN) is an extension of a stochastic petri net (SPN). A rigorous mathematical description of stochastic reward nets is there in Muppala et al. [15].

Petri net in its original definition suffers from the problem of state-space explosion. So, over the time various features such as guard, priority relationship, and inhibitor arcs have been added to PNs to provide a concise description of a given system.

Associating exponentially distributed firing times with the transitions of the PN results in a Stochastic Petri net (Molly [11]). Allowing transition to have either zero firing times (immediate transitions) or exponentially distributed firing times (timed transitions) gives rise to the Generalized Stochastic Petri Net (GSPN) (Ajmone-Marson et al. [17, 18]) as already seen.

By associating reward rates with the markings of the SPN, SRN is obtained. As SRN can be automatically converted into a Markov reward model thus permitting the evaluation of not only performance and availability but also their combination. Putting all this together, SRN can formally be defined as:

SRN: A marked SRN is a tuple $A = (P, T, DI, DO, DH, \widehat{G}, >, \lambda, PS, M_0, r)$ [19] where:

$P = p_1, p_2, \ldots, p_N$ is a finite set of places

$T = t_1, t_2, \ldots, t_M$ is a finite set of transitions

$\forall p_i \in P, \forall t_j \in T, DI_{ij} : IN^N \rightarrow IN$ is the marking dependent multiplicity of the input arc from place p_i to transition t_j; if the multiplicity is zero, the input arc is absent

$\forall p_i \in P, \forall t_j \in T, DO_{ij} : IN^N \rightarrow IN$ is the marking dependent multiplicity of the output arc from transition t_j to place p_i; if the multiplicity is zero, the output arc is absent

$\forall p_i \in P, \forall t_j \in T, DH_{ij} : IN^N \rightarrow IN$ is the marking dependent multiplicity of the inhibitor arc from place p_i to transition t_j; if the multiplicity is zero, the inhibitor arc is absent

$\forall t_j \in T, \hat{G}_j : IN^N \rightarrow \{0, 1\}$ is the marking dependent guard of the transition t_j

$>$ is a transitive and irreflexive relation imposing a priority among transitions. In a marking M_j, t_1 is enabled iff it satisfies its input and inhibitor conditions, its guard evaluates to 1, and no other transition t_2 exists such that $t_2 > t_1$, and t_2 satisfies all other conditions for enabling

$\forall t_j \in T$ such that t_j is a timed transition, $\lambda_j : IN^N \rightarrow IR^+$ is the marking dependent firing rate of transition t_j and $\lambda = [\lambda_j]$

$\forall t_j \in T$ such that t_j is an immediate transition, $PS_{t_j} : IN^N \rightarrow [0, 1]$ is the marking dependent firing probability for transition t_j, given that the transition is enabled.

$M_0 \in IN^N$ is the initial marking

$r_j \in IR$ is a reward rate associated with each tangible marking M_j that is reachable from the initial marking M_0, and $r = [r_j]$

3.6.6 Deterministic and Stochastic Petri Net (DSPN)

Stochastic Petri nets (SPNs) are well suited for model-based performance and dependability evaluation. Most commonly, the firing times for the transitions are exponentially distributed, leading to an underlying CTMC (continuous time Markov chain). In order to increase modeling power, several classes of non-Markovian SPNs were defined, in which the transition may fire after a non-exponentially distributed firing time.

A particular case of non-Markovian SPNs is the class of deterministic and stochastic Petri nets (DSPNs) [19], which allows transitions to have deterministic firing times along with transition with exponential firing times. A DSPN with restriction that at any time at most one deterministic transition may be enabled. When this condition is met, it has been shown that the marking process corresponds to Markov regenerative process [6, 7]. Being a non-Markovian system, analysis method popular for solving DSPNs are bases on Supplementary variable and imbedded Markov chain [1]. Stationary analysis method for DSPNs with mentioned conditions are presented in [20] and citemarsanon and the transient analysis are addressed in [6].

3.6.7 Queueing Networks

Queueing networks are a widely used performance analysis technique for those systems, which can be naturally represented as networks of queues. Systems, which have been successfully with queueing networks, include computer systems, communication networks, and flexible manufacturing systems [4].

A queueing system consists of three types of components:

1. Service centers: a service center consists of one or more queues and one or more servers. The server represent the resources of the system available to service customers. An arriving customer will immediately be served if a free server can be allocated to the customer or if a customer in service is preempted. Otherwise, the customer must wait in one of the queues, until a server become available.
2. Customers: a customer is one which demand service from the service centers and which represent load on the system.
3. Routes: routes are the paths which workloads follow through a network of service centers. The routing of customers may be dependent on the state of the network. If the routing is such that no customers may enter or leave the system,

the system is said to be closed. If the customers arrive externally and eventually depart, the system is said to be open. If some classes of customers are closed and some are open, then the system is said to be mixed.

To complete specify the queueing network following parameters are defined:

- The number of service centers
- The number of queues at each service center. For each of these queues, we further need to define:
  - The capacity of each queue, which may be of finite capacity k or infinite.
  - The queue scheduling discipline, which determines the order of customer service. Different customer classes may have different scheduling priorities. Common scheduling rules include First-Come First-Serve (FCFS), Last-Come First Serve, highest priority first with or without preemption, round robin (RR) and processor sharing (PS)
  - For open classes of customers, we need to define an input source distribution of each customer class at each queue. (this distribution is usually given by an exponential distribution with parameter λ)
- The number of servers at each service center. For each of these servers we further need to define:
  - The service time distribution for each customer class at each server. (this is usually exponential with parameter μ)
- The routing probability matrix for each customer class. This matrix specifies the probabilistic routing of customers between service centers, with the (i, j)th element giving the probability that a customer leaving service center i will proceed to service center j. These transition are assumed to be instantaneous.

Queueing networks can be mapped into CTMC or DTMC, which ever may be appropriate and can be analysed.

A certain class of networks which satisfy reversibility can be efficiently analysed using so called product-form solution techniques, the two most well-known of which are Mean-Value Analysis (MVA) and the convolution method.

Queueing networks are widely used because they are often easy to define, parameterise and evaluate. However, they lack of facilities to describe synchronization mechanism and difficult to solve under varying scheduling policies.

Queueing networks have been successfully used in performance modeling of computer and communication systems [20]. They are especially suited for representing resource contention and queueing for service. Most of the analysis technique so far have concentrated on the evaluation of averages of various performance measures like throughput, utilization and response time using efficient algorithms such as convolution and mean value analysis (MVA). For real-time systems, however, the knowledge of response time distributions is required in order to compute and/or minimize the probability of missing a deadline.

3.6.8 Stochastic Process Algebra (SPA)

A process algebra (PA) is an abstract language which differs from the formalisms we have considered so far because it is not based on a notion of flow. Instead, systems are modeled as a collection of cooperating agents or processes which execute atomic actions. These actions can be carried out independently or can be synchronized with the actions of other agents.

Since models are typically built from smaller components using a small set of combinators, process algebras are particularly suited to the modeling of large systems with hierarchical structure. This support for compositionality is complemented by mechanisms to provide abstraction and compositional reasoning.

Widely known process algebras are Hoare's Communicating Sequential Process (CSP) and Miner's Calculus of Communicating Systems. These algebras do not include notion of time so they can only be used to determine qualitative correctness properties of the system such as free from racing, deadlock and livelock. Stochastic Process Algebras (SPAs) additionally allow for quantitative performance/reliability analysis by associating a random variable, representing duration, with action/state. Several tools have been developed for SPA, such as PEPA, TIPP, MPA, SPADES and EMPA.

We will describe SPA, using Markovian SPA PEPA. PEPA models are built from components which perform activities of form (α, r); where α is the action type and $r \in \Re^{+} \cup T$ is the exponentially distributed rate of the action. The special symbol t denotes an passive activity that may only take place in synchrony with another action whose rate is specified [21].

Interaction between components is expressed using a small set of combinators, which are briefly described below [21]:

Sequential composition: Given a process $P, (\alpha, r). P$ represents a process that performs an activity of type α, which has a duration exponentially distributed with mean $1/r$, and then evolves into P.

Constant: Given a process $Q, P = Q$ means that P is process which behaves in exactly the same way as Q.

Selection: Given processes P and Q, $P + Q$ represents a process that behaves either as P or as Q. The current activities of both P and Q are enabled and a race condition determines into which component the process will evolve.

Synchronization: Given processes P and Q and a set of action types $L, P \rhd \lhd Q$ defines the concurrent synchronized execution of P and Q over the cooperation set L. No synchronization takes place for any activity $\alpha \notin L$, so that activities can take place independently. However, an activity $\alpha \in L$ only occurs when both P and Q are capable of performing the action. The rate at which the action occurs is given by the minimum of the rates at which the two components would have executed the action in isolation.

Cooperation over the empty set $P \rhd \lhd Q$ represents the independent concurrent execution of processes P and Q and is denoted as $P||Q$.

Encapsulation: Given process P and a set of actions $L, P/L$ represents a process that behaves like P except that activities $\alpha \in L$ are hidden and performed as a silent activity. Such activities cannot be part of a cooperation set.

PEPA specifications can be mapped onto continuous time Markov chains in a straightforward manner. Based on the labeled transition system semantics that are normally specified for a process algebra system, a transition diagram or derivation graph can be associated with any language expression. This graph describe all possible evolutions of a component and, like a reachability graph in the context of GSPNs, is isomorphic to a CTMC which can be solved for its steady-state distribution.

The main advantage of process algebras over other formalisms is their support for compositionality, i.e. the ability to construct complex models in a stepwise fashion from smaller building blocks, and abstraction, which provides a way to treat components as black boxes, making their internal structure invisible. However, unlike the other formalisms we have considered, process algebras lack an intuitive graphical notion so does not always present a clear image of the dynamic behavior of the model.

3.7 Tools

3.7.1 SPNP

Stochastic Petri Net Package (SPNP) [22] is a versatile modeling tool for performance, dependability and performability analysis of complex systems. Input models based on theory of stochastic reward nets are solved by efficient and numerically stable algorithms. Steady-state, transient, cumulative transient, time-averaged and up-to-absorption measures can be computed. Parametric sensitivity analysis of these measures is possible. Some degree of logical analysis capabilities are also available in the form of assertion checking and the number and types of markings in the reachability graph. Advanced constructs, such as marking dependent arc multiplicities, guards, arrays of places and transitions, are available. The modeling complexities can be reduced with these advanced constructs. The most powerful feature of SPNP is the ability to assign reward rates at the net level and subsequently compute the desired measures of the system being modeled. SPNP Version 6.0 has the capability to specify non-Markovian SPNs and Fluid Stochastic Petri Nets (FSPNPs). Such SPN are solved using discrete-event simulation rather than by analytical-numeric methods. Several types of simulation methods are available: standard discrete-event simulation with independent replications or batches, importance splitting techniques (splitting and Restart), importance sampling, regenerative simulation without or with importance sampling, thinning with independent replications, batches or importance sampling.

3.7.2 TimeNET

TimeNet [23, 24] is a graphical and interactive toolkit for modeling with stochastic Petri nets. TimeNET has been developed at the Institut fur Technische Informatik of the Technische Universitat Berlin, Germany. It provides a unified framework for modeling and performance evaluation of non-Markovian stochastic Petri nets. It uses a refined numerical solution algorithm for steady-state evaluation of DSPNs with only one deterministic transition enable in any marking. Ex-polynomial distributed firing times are allowed for transitions. Different solution algorithms can be used, depending on the net class. If the transition with non-exponential distributed firing times are mutually exclusive, TimeNET can compute steady-state solution. DSPNs with more than one enabled deterministic transition in a marking are called concurrent DSPNs. TimeNET provides an approximate analysis technique for this class. If the mentioned restrictions are violated or the reachability graph is too complex for a model, an efficient simulation component is available. A master/slave concept with parallel replications and techniques for monitoring the statistical accuracy as well as reducing the simulation length in case of rare events are applied. Analysis, approximation, and simulation can be performed to the same model classes. For more details refer to TimeNET user manual [24].

3.8 Summary

Real life system are usually complex, to model then a family random variables is required. These family of random variables are termed as stochastic variables. These stochastic variables have some unique characteristic. Dealing at lower level, i.e. state-transition, for a complex problem becomes unmanageable. So, higher level modeling formalisms are required. This is the theme of this chapter.

References

1. Cox DR, Miller HD (1970) The theory of stochastic processes. Methuen, London
2. Trivedi KS (1982) Probability statistics with reliability, queueing, and computer science applications. Wiley, New York
3. Xinyu Z. (1999) Dependability modeling of computer systems and networks. Ph.D. thesis, Department of Electrical and Computer Engineering, Duke University
4. IEC 60880-2.0: Nuclear power plants—instrumentation and control systems important to safety—software aspects for computer-based systems performing category a functions, 2006
5. Erhan C (1975) Introduction to stochastic processes. Prentice-Hall, Englewood Cliffs
6. Choi H, Kulkarni VG, Trivedi KS (1993) Transient analysis of deterministic and stochastic petri nets. In: Proceedings of the14th international conference on application and theory of petri nets, pp 166–185

7. Choi H, Kulkarni VG, Trivedi KS (1994) Markov regenerative stochastic petri nets. Perform Eval 20:337–357
8. Meyer JF (1980) On evaluating the performability of degradable computing systems. IEEE Trans Comp C 29(8):720–731
9. Meyer JF (1982) Closed-form solutions of performability. IEEE Trans Comp C 31(7):648–657
10. Puliafito A, Telek M, Trivedi KS (1997) The evolution of stochastic petri nets. In: Proceedings of World Congress of Systems and Simulation, WCSS 97:97
11. Molloy MK (1982) Performance analysis using stochastic petri nets. IEEE Trans Comp C 31(9):913–917
12. Murata T (1989) Petri nets: properties, analysis and applications. Proceedings IEEE 77(4):541–580
13. Peterson PL (1977) Petri nets. ACM Comput Surv 9(3)
14. Peterson PL (1981) Petri net theory and modeling of systems. PHI, Englewood Cliffs
15. Jogesh M, Gianfranco C, Trivedi KS (1994) Stochastic reward nets for reliability prediction. Commun Reliab Maintainabil Serviceabil 1(2):9–20
16. Marsan MA, Balbo G, Conte G (1984) A class of generalized stochastic Petri nets for the performance evaluation of multiprocessor systems. ACM Trans Comp Syst 93:93–122
17. Ajmone Marsan M, Balbo G, Conte G (1984) A class of generalized stochastic petri nets for the performance evaluation of multiprocessor systems. ACM Trans Comp Syst 2(2):93–122
18. Marson MA, Balbo G, Bobbio A, Chiola G, Conte G, Cumani A (1989) The effect of execution policies on the semantics and analysis of stochastic petri nets. IEEE Trans Softw Eng 15(7):832–846
19. Bukowski JV (2001) Modeling and analyzing the effects of periodic inspection on the performance of safety-critical systems. IEEE Trans Reliabil 50(3):321–329
20. Marsan MA, Chiola G (1987) On petri nets with deterministic and exponentially distributed firing times. In: Advances in Petri Nets 1986, Lecture Notes in Computer Science 266, pp 132–145
21. Diaz JL, Lopez JM, Gracia DF (2002) Probabilistic analysis of the response time in a real time system. In: Proceedings of the 1st CARTS workshop on advanced real-time technologies, October
22. Trivedi KS (2001) SPNP user's manual. version 6.0. Technical report
23. Zimmermann A (2001) TimeNET 3.0 user manual
24. Zimmermann A, Michael Knoke (2007) Time NET 4.0 user manual. Technical report, August

Chapter 4
Dependability Models of Computer-Based Systems

4.1 Introduction

Computer-based reactive systems which interact with their environment in a timely manner are called *real-time* systems. The main characteristics of real-time systems which distinguishes them from others is that the correctness (or healthiness) of the system depends not only on the *value* of its response, but also on the *time* at which it is produced. So, in real-time systems two kinds of hazardous faults are recognized: (i) incorrect response (*value* failure) and, (ii) deadline miss (*timeliness* failure).

Computer-based systems (CBS) which continuously interact with their environment and try to keep some parameters of environment under pre-defined limits are called *control* systems. The main characteristic of control system is *stability*, and *quality of performance*. So, a control system is said to be failed if it becomes unstable or its quality of performance deteriorates than acceptable.

When these systems are used in critical applications, failure of these may cause loss of life, damage to environment and/or huge investment (or economic) loss. These systems are also referred as critical systems. These critical systems can be categorized as, (i) *safety-critical*, (ii) *mission-critical* and, (iii) *economically-critical* systems based on type or extent of loss/damage. Examples of safety-critical CBS are shutdown system of nuclear reactor, digital flight control computer of aircraft, braking system of wire-by-brake system of a automobile etc. Control & coding unit (CCU) of an avionic system, navigation system of a spacecraft and navigation system of a guided missile etc., are example of mission-critical systems. Economically-critical systems include reactor control system of a nuclear reactor, fuel-injection system of automobile etc.

In this chapter, dependability attributes applicable to above critical systems are discussed. Methods and models available in literature for these attributes are also discussed.

A. K. Verma et al., *Dependability of Networked Computer-based Systems*, Springer Series in Reliability Engineering, DOI: 10.1007/978-0-85729-318-3_4,

4.2 Dependability Attributes

Dependability of computing system is defined by Algirdas Avizienis et al. [1] as "ability to deliver service that can justifiably be trusted". The service delivered by a system is its behavior as it is perceived by its user(s). User could be another system (physical, human) that interacts with the former. Service is delivered when the service implements the system function, where function is the behavior of the system described by its specification.

A system failure is an event that occurs when the delivered service deviates from correct service. A failure is a transition from correct service to incorrect service. Failure is manifestation of error, which in turn is caused by fault [1–3]. Based on domain, faults are categorized as *physical* and *information* faults.

As discussed earlier, a computer-based system when used in critical applications, can be categorized as *safety-critical*, *mission-critical* and *economically critical*. Dependability attributes applicable to these systems are safety, reliability and availability, respectively. Figure 4.1 shows the applicable dependability attributes pictorially.

Definition 4.1 Safety-critical systems: systems required to ensure safety of equipment under control (EUC), people and environment.

Definition 4.2 Mission-critical systems: systems whose failure results in failure/ loss of mission.

Definition 4.3 Economically-critical systems: systems whose failure result in availability of EUC, causing massive loss of revenue.

In next section a review of dependability attributes—reliability, availability and safety—are presented. Dependability attribute, safety has been discussed with great emphasis on new safety model incorporating demand rate.

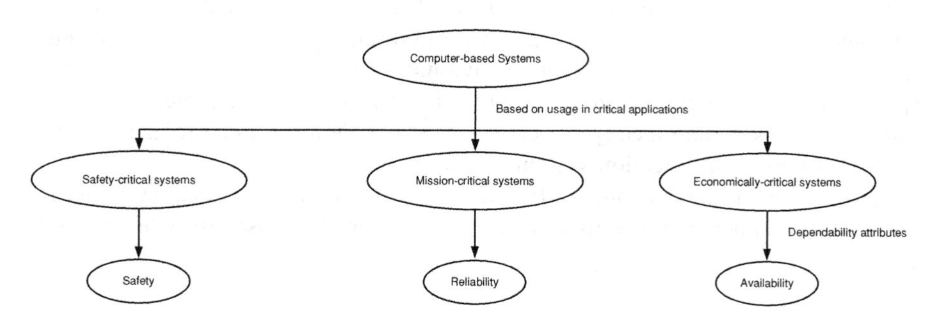

Fig. 4.1 Failure domains and dependability attributes

4.3 Reliability Models

In case of computer-based systems, component or hardware faults and software faults are the causes of system failure. CBS are implemented using electronic components. To estimate hazard rate (λ) of electronic components two accepted techniques are there, (i) accelerated life testing, (ii) component reliability models. These failure models are based on failure data of components. Accepted models are MIL-HDBK-217F, Telcordia TR/SR 332, British Telecom HRD4 and HRD5 etc. These models give hazard rate based on operating condition and various stress factors.

Electronic components do not have moving parts, so fatigue failures are not there. Failure can be assumed random and system posses no time memory, i.e. age. So, once hazard rate is available exponential distribution can be used for mathematical representation of component reliability with time.

4.3.1 Combinatorial Models

Combinatorial models are a class of reliability models that represent system failure behaviors in terms of combination of component failures [4, 5]. Because of their concise representation of system failure, combinatorial models have long been used for reliability analysis. Reliability block diagrams, fault trees and reliability graphs are three major types of combinatorial models. A brief description of these is given below.

4.3.1.1 Reliability Block Diagrams

RBD is a graphical tool consisting of blocks. Individual blocks may represent single component, module, sub-system, and/or logical blocks etc. The blocks are connected in a manner to depict the reliability-wise relationship among the blocks. This is the reason RBD is also termed as structure-oriented method. For example, pair of shoes is reliability-wise in series, while a pair of eyes is in parallel. For the system to be successful in its operation at least one path must exist between input block and output block. The RBD can be analyzed using analytical or simulation methods. Analytical methods include, series-parallel method, MooN (M out-of N) and Bayes Method. Simulation method includes Monte-Carlo simulation. For complex system with large number blocks and/or complex interaction among blocks simulation is used.

4.3.1.2 Reliability Graphs

A reliability graph model consists of a set of nodes and edges, where the edges represent components that can fail or structural relationship between the components. The graph contains nodes, a node with no incoming edges is called *source*, while a node with no outgoing edges is termed as *sink*. Reliability graphs are similar to RBD and comes under the broad category of structure-oriented models. Reliability graphs are mostly suited for complex system where reliability-wise relationship among bocks is more complicated then series-parallel. Graph theoretic methods—cut-set, path-set and BDD [4]—be used for solution.

4.3.1.3 Fault Trees

A fault tree is a graphical representation of the combination of events that can cause the occurrence of overall desired event, e.g. system failure in case of reliability modeling. RBD and RG can only model hardware failures, while fault tree can model hardware failures as well as failures on account of software failure, human errors, operation and maintenance errors and environment influences on the system.

Fault tree identifies relationships between an undesired system event and the subsystem failure events that may contribute to its occurrence. Fault tree development employs a top-down approach, descending from the system level to more detailed levels of subsystems and component levels. It is well suited to evaluate the reliability considerations at each stage of the system design.

Fault tree provides both qualitative or quantitative system reliability. Environmental and other external influences can easily be considered in fault tree analysis. It provides visual and graphical aid to the analyst.

4.3.2 Dynamic Models

Combinatorial models are the simplest and widely used methods for reliability modeling. These methods are not suitable for modeling where system failure depends upon sequence of failure occurrence. Combinatorial models give point estimate of reliability, i.e. for a given scenario. While system may degrade with time, components may fail and could be repaired and restored back. This repair could be perfect or imperfect.

To include sequence of failure fault trees are extended to dynamic fault tree. Markov models are suitable to model sequence of failure, degradation with time, failure and repair, and reliability with time. Brief description of these two are given below:

4.3.2.1 Dynamic Fault Trees

Fault tree discussed above has a major shortcoming—inability to capture sequence dependencies [6]. For example consider a dynamic redundant system with one active component and one standby spare connected with a switch unit. If the switch unit fails before the active component fails, then the standby unit cannot be switched into active operation and the system fails when the active components fails. Thus, the failure criteria depends not only on the combination of events, but also on their sequence.

Dynamic fault tree tries to eliminate this limitation—sequence of failure—of fault tree by incorporating functional dependency and priority gates.

4.3.2.2 Markov Models

Markov chains are widely used for modeling and analyzing problems of stochastic nature. A stochastic process is a Markov process if its future evolution depends on its current state only. Means, the next state of the process is independent of the history of the process, i.e. it only depends upon its current state. The processes of this nature are termed as Markov process. Equation 4.1 describe a Markov Process [7, 8].

$$\Pr\{X(t_n = j|X(t_{n-1}) = i_{n-1}, \ldots, X(t_0) = i_0\} = \Pr\{X(t_n) = j|X(t_{n-1}) = i_{n-1}\} \tag{4.1}$$

Whether a particular system leads to a Markov process depends on how the random variables specifying the stochastic process are defined. For example, consider a component, such as a IC, which may fail. Let the component be checked periodically and classified as being in one of three states, (i) satisfactory, (ii) unsatisfactory and, (iii) failed. Let these three states are termed as state 0, 1 and 2 respectively. The process has been depicted in Fig 4.2. The transition probabilities of this example is given as:

$$P = \begin{matrix} & \begin{matrix} 0 & 1 & 2 \end{matrix} \\ \begin{matrix} 0 \\ 1 \\ 2 \end{matrix} & \begin{bmatrix} p_{00} & p_{01} & p_{02} \\ 0 & p_{11} & p_{12} \\ 0 & 0 & 1 \end{bmatrix} \end{matrix} \tag{4.2}$$

In (4.2) rows correspond to initial state while column to final state. Transition matrix element p_{00} depicts the probability of remaining in state 0, while p_{01} depicts probability of transition to state 1 from state 0. Using this transition matrix, next state probabilities can be estimated using following relation:

$$p^{n+1} = p^n P \tag{4.3}$$

where p is the state probability; P is the transition matrix.

From (4.3), it is evident that component state at any instant $n + 1$ is dependent only on the previous state at instant n, and these two states are related to each other by transition matrix. So, this component state problem in current form is modeled by a Markov process.

Now, modify the problem statement, if state 1 (unsatisfactory) is entered, the component remain in that state for exactly two time periods before passing to state 2.

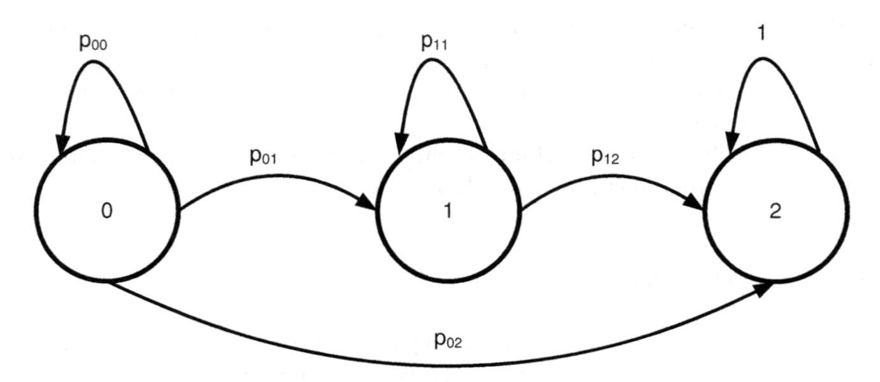

Fig. 4.2 Discrete time Markov model of component example

From this modified problem definition, it is evident that system has memory in state 1, and by definition it cannot be modeled directly as Markov process. New process is described as:

$$\begin{aligned} prob(p^{n+1} &= 1 | p^n = 1, p^{n-1} = 0) = 1 \\ prob(p^{n+1} &= 1 | p^n = 1, p^{n-1} = 1) = 0 \end{aligned} \tag{4.4}$$

But a simple extension to state space of this problem converts the problem into Markov process. The process involves dividing the original state 1 into two states, (1,0) and (1,1), where (1,0) is the state corresponding to $p^n = 1$, $p^{n-1} = 0$ and (1,1) the state corresponding to $p^n = 1$, $p^{n-1} = 1$. The new process with four states has transition probability matrix as:

$$P = \begin{array}{c} \\ 0 \\ (1,0) \\ (1,1) \\ 2 \end{array} \begin{array}{c} \begin{array}{cccc} 0 & (1,0) & (1,1) & 2 \end{array} \\ \begin{bmatrix} p_{00} & p_{01} & 0 & p_{02} \\ 0 & 0 & 1 & 0 \\ 0 & 0 & 0 & 1 \\ 0 & 0 & 0 & 1 \end{bmatrix} \end{array} \tag{4.5}$$

The addition of this new state enables the problem to modeled as Markov process.

A Markov process is characterized by its states and transitions. Time to transition and states may be discrete or continuous, independent of each other. So a Markov process can further be divided into four domains. A Markov process with discrete states and continuous time to transition is termed as "Continuous Time Markov Chain" (CTMC), while process with discrete state and discrete time is termed as "Discrete Time Markov Chain" (DTMC). CTMC is well suited for reliability analysis of electronic equipments, as their failure process is characterized as Markov. In reliability analysis states of the CTMC depicts the state of the system and transition depicts the failure rate. CTMCs are capable of taking redundancy and repair activity into account.

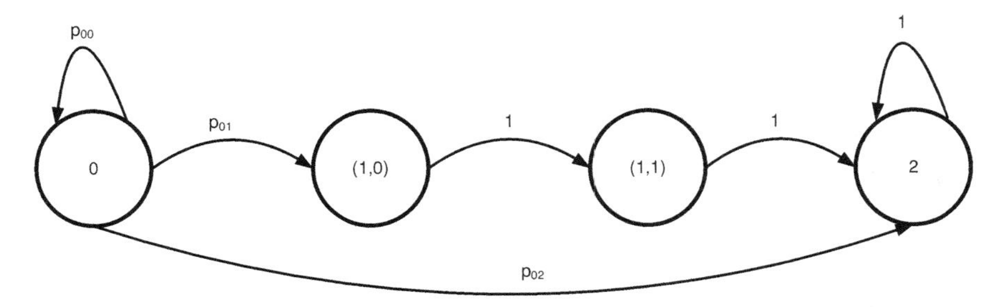

Fig. 4.3 Discrete time Markov model of component example

4.3.3 Software Reliability

Software is an integral part of any computer-based system. Software can cause system failure. Major causes of system failure due to software, as per literature are as follows:

- Specification
- Design
- Interaction
- Stress

In literature, unlike electronic components, there is no widely accepted, standardized method for software reliability prediction. A number of methods has been proposed by contemporary researchers. Various models of software worthiness estimation and growth—failure rate models, reliability models and reliability growth models—are available in literature. Some of them are given as below [9–11]:

1. *Failure Rate models* these are based on modeling the software failure intensity from software test data.

   (a) Jelinski–Moranda model
   (b) Schick–Walvertom model
   (c) Jelinski–Moranda Geometric model
   (d) Goel–Okumoto debugging model [12]

2. *NHPP software Reliability model* it assumes faults are dormant, and time to uncover follows non-homogeneous Poisson process.
3. *State based models* these models use control flow graph of the software to represent the architecture of the system, which could be modeled as DTMC, CTMC or SMP (semi-Markov process).

   (a) Littlewood model [13]
   (b) Cheung model [14]
   (c) Laprie model [15]

(d) Kubat model [16]
(e) Gokhale et al. model [17, 18]
(f) Ledoux model [19]
(g) Gokhale et al. reliability simulation approach

4. *Path-based approach* Here the software reliability is computed considering the possible execution paths of the program.

   (a) Shooman model [20]
   (b) Krishnamuthy and Mathur model
   (c) Yacoub model [21]

5. *Additive models* these models estimate the system failure intensity as sum of component failure intensities under the assumption that individual component reliabilities can be modeled by NHPP.

   (a) Xie and Wohlin model [22]
   (b) Everett model [23]

In this text, software reliability of CBS is assumed to be known a priori.

4.4 Availability Models

Availability is a measure for system which are subjected to failure and repair. Availability refers to fraction of time system spends in *UP* state. Mathematically it is described as:

$$\text{availability upto time } t = A(t) = \frac{\text{Total time spent in } UP}{\text{Total time `}t\text{'}} \tag{4.6}$$

From (4.6), it is evident that once system history (time spent in *UP* and *DN* state) is available, availability can be determined.

$$A(t) = \frac{t_{UP}}{t_{UP} + t_{DN}} \tag{4.7}$$

Figure 4.4 gives a typical trace of system state with time. From this, statistically mean value of t_{UP} and t_{DN} can be determined. With this, availability seems to be a posteriori measure of system dependability. A estimate of being in *UP* state for a given duration may give availability estimate a priori. For computer-based system, Markov models are widely used to estimate the time a system spend in *UP* state. For a repairable system *MTBF* and *MTTR* can be found from system history or estimate using model, in that case availability is given as:

$$A = \frac{MTBF}{MTBF + MTTR} \tag{4.8}$$

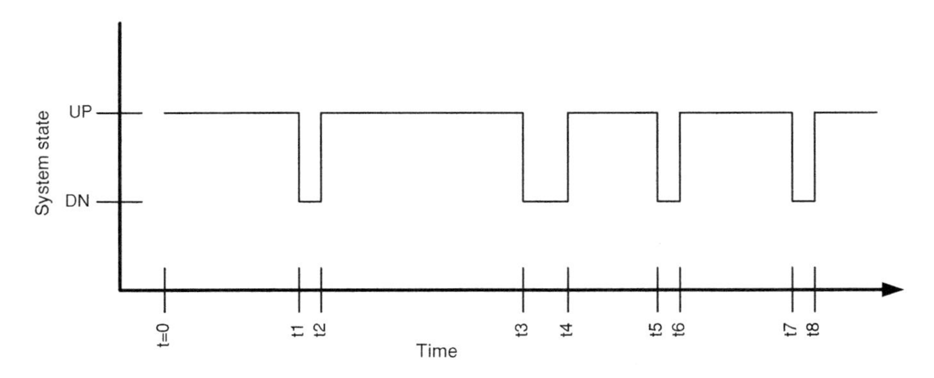

Fig. 4.4 System state *UP*, *DN* history plot

4.5 Safety Models

Safety-critical systems are used for automatic shutdown of EUC; whenever the equipment or plant parameters go beyond the acceptable limits for more than acceptable time. These kinds of systems are used in a variety of industries; such as oil refining, nuclear power plant, chemical and pharmaceutical manufacturing etc. When the safety system is functioning correctly (successfully), it permits the EUC to continue provided its parameters remain within safe limits. If the parameters move outside of an acceptable operating range for a specified time, the safety system automatically shutdowns the EUC in a safe manner.

The safety systems generally have some redundancy and can tolerate some failures while continuing to operate successfully. As discussed in ref. [24–27] system's independent channels can fail leading system to following states:

1. Safe failure (*SF*) state where it erroneously commands to shutdown a properly operating equipment. Taking a channel off line and shut-down of a channel is also referred as safe failure.
2. Fail Dangerous Detected (*DD*) state where channel(s) is (are) failed in dangerous mode, but detected by internal diagnostics, and announced.
3. Fail Dangerous Undetected (*DU*) state where channel(s) is failed in dangerous mode and not detected by internal diagnostics, hence not announced.

The safety system can fail in distinctly two different ways [24–28]

1. Safe Failure (F_S), failure which does not have potential to put the safety system in a hazardous or fail-to-function state [24]. This occurs when more than tolerable numbers of channels are in safe failure. This type of failure is referred to in a variety of ways including fail safe [25, 28], false trip and false alarm.
2. Dangerous Failure (*DF*), failure which has the potential to put the safety system in a hazardous or fail-to-function state [24]. More than tolerable number of

channel in DD and/or DU lead to this failure. The system fails in such a way that it is unable to shutdown the EUC properly when shutdown is required (or demanded).

Dangerous failures are important from safety point of view. A survey of recent research work related to safety quantification indicates that there are diverse safety indices, methods and assumptions about the safety systems. Safety indices used are *PFD* (probability of failure on demand) [24, 26, 27, 29–33], $MTTF_D$ (mean time to dangerous failure) [25, 28], $MTTF_{sys}$ (mean time to system failure) [34], *MTTUF* (mean time to unsafe failure) & S_{SS} (steady state safety) [35], and *MTTHE* (mean time to hazardous event) [36]. Simplified equations [24, 26, 29, 32, 33], Markov model [25, 27, 28, 30, 31, 34–37] and fault tree [33] are the methods used for safety quantification. Safety indices of [35, 36] consider only repair. Bukowski [25] considers repair as well as periodic inspection to uncover undetected faults. Refs. [24, 26, 27, 29, 32, 33] consider common cause failures (CCF), periodic inspection along with repair, and [37] consider demand rate. Ref. [32] discusses the CCF model (β factor) of [24] and suggests generalization, multiple beta factor (MBF) (multiple beta factor).

4.5.1 Modeling of Common Cause Failures (CCF)

4.5.1.1 β-Factor Model

β-factor model is very simple method to model common cause failure (CCF). The problem with β-factor model is that it has no distinction between different voting logics. To overcome this problem, different β's, based on heuristics are used for different voting logics.

4.5.1.2 Multiple Beta Factor (MBF) for Common Cause Failures (CCF)

There exists some apparent inconsistency or ambiguity regarding the definition and use of terms *random hardware failures* and *systematic failures*, and the way these are related to common cause failures (CCFs) [32]. In this section this classification is discussed and some suggestions are outlined.

IEC 61508 uses β-factor model for CCF. β-factor model does not distinguishes between performance of various voting logics like 1*oo*2 and 2*oo*3. Before proceeding further, it is better to have a look at the failure classification of IEC 61508. As per standard, failures of SIS (safety instrumented system) can be categorized either as a random hardware failure or as a systematic failure. Where random hardware failure means failure that occurs without the failed component being exposed to any kind of 'excessive' stress [32]. While systematic failure is related in deterministic way to a certain cause, which can be eliminated by improving/modifying design, manufacturing process, operational procedure etc. [32]. So, it includes all types of failures caused by design errors.

Most safety standards makes a clear distinction between these two failures categories, and quantify only random hardware failures only. IEC 61508 in context of CCF describes: "However, some failures, i.e. common cause failures, which result from a single cause, may effect more than one channel. These may result from a systematic fault (for example, a design or specification mistake) or an external stress leading to an early random hardware failure." Therefore, the CCFs may either result from a systematic fault or it is random hardware failure due to common excessive stress on the components. While these standards model CCFs arising from excessive stresses on the hardware are quantified. Hoskstad et al. [32] present a detailed failure classification by cause of failure, it is shown in Fig. 4.5.

As shown in the figure, random failures are physical failures. The main causes of random failures are time (age) and stress. On the other hand, systematic failures are non-physical and its causes of introduction are design and interactions. Design includes mistakes made in specifications, during engineering and construction. Interaction mainly deals with man-made errors during operation or maintenance/ testing.

Random hardware failures are detected by deviation of performance or service from specified due to physical degradation. Due to systematic failure, the delivered service or performance deviates from the specified, without a random hardware failure being present. Safety standards such as IEC 61508, does not quantify systematic failures. But it may be advantageous to introduce various measures for loss of safety, arising from different contributions.

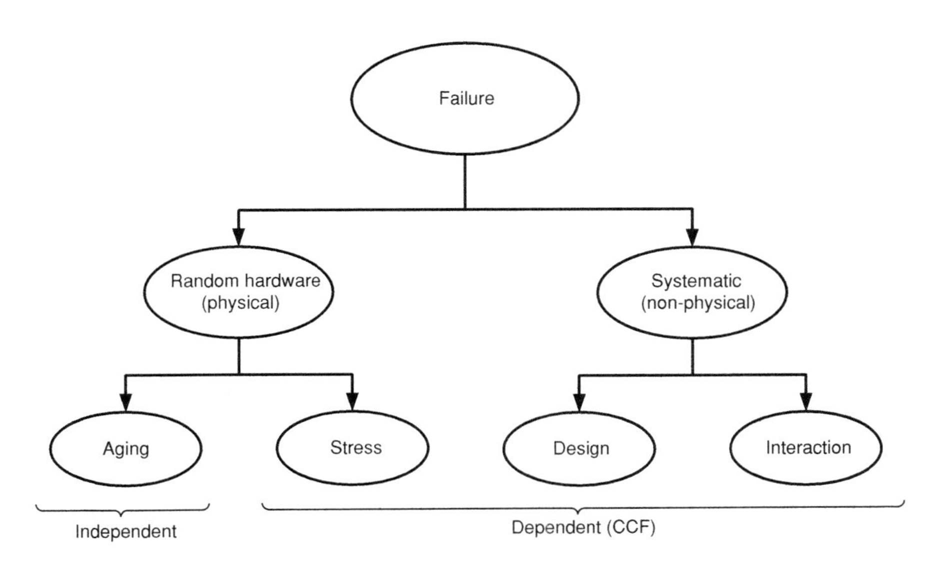

Fig. 4.5 Failure classification by cause of failure [32]

4.5.2 Safety Model of IEC 61508

Safety index *PFD* [24] has already been published as a standard. As per IEC 61508 [24], a typical trace of system states is given in Fig. 4.6. The mean probability of being in *DD* or *DU* state is *PFD.* In the figure times marked as t_{Di} denotes occurrence of *i*th detected dangerous failure, t_{Rj} completion of *j*th repair, t_{Uk} occurrence of *k*th undetected dangerous failure, and t_{pl} time of *l*th proof-test.

IEC 61508 [24] gives simplified equations for safety evaluation. Since the inception of IEC 61508 [24], its concepts and methods for loss of safety have been made clearer and substantiated by Markov models. A review of different techniques by Rouvroye [38] suggests Markov analysis covers most aspects for quantitative safety evaluation. Bukowski [30] also compares various techniques for *PFD* evaluation and defends Markov models. Zhang [27] provides Markov model for *PFD* evaluation without considering demand rate and modeling imperfect proof-tests.

Demand refers to a condition when the safety system must shut down EUC. The condition arises when EUC parameters move outside of an acceptable operating range for a specified time. A trace of system state considering demand is shown in Fig. 4.7. Downward arrows at Demand Incidence line shows the time epochs of demand arrival. Epochs marked as $\circ$ at *DEUC* line, denote successful action taken by safety system at demand arrival. While epoch marked as $\times$, denotes unsafe failure of the safety system and damage to EUC.

Bukowski [37] proposes a Markov model based safety model, similar to *PFD,* considering demand rate. This model does not consider periodic proof-tests. Detailed comparison of these models with the one proposed here is given in Sect. 4.5.4.2.

System model developed here is similar to the model of IEC 61508 [24]. It uses Markov model for analysis. This model explicitly consider periodic proof-test (perfect or imperfect), demand rate and safe failures. Incorporation of safe failure

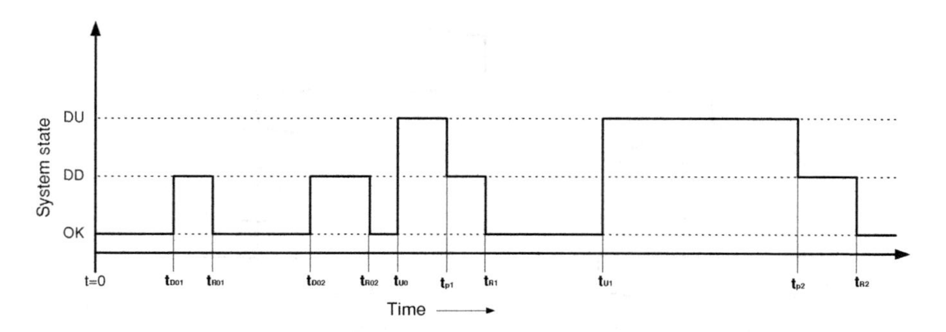

Fig. 4.6 A typical trace of system states as per IEC 61508 [24]. System may make transitions to states *DD* and *DU*, based on the type of failure. System can be restored back from *DD* state by means of repair, after the failure is detected. *DU* state can be detected only during proof-test. So, a system will remain in *DU* state till proof-tests. Safety measure of IEC 61508 [24], *PFD*, gives the mean probability of finding the system in state *DD* or *DU*

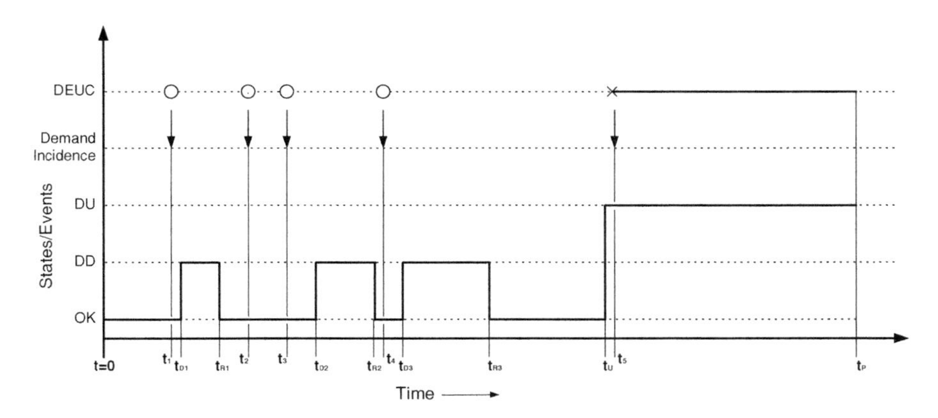

Fig. 4.7 A typical trace of system states considering demands. This trace is same as Fig. 4.6, with addition of demand arrival epoch and state *DEUC*. The safety system will damage EUC, if it is in *DD* or *DU* state, at demand arrival. Safety measure considering demand is the probability of reaching state, *DEUC*

enables modeling of all possible system states and estimation of additional measures such as availability (or probability of being in one or more specified states) for a specified amount of time.

System description and assumptions about the system to derive Markov model is given in next section.

4.5.3 System Model

The computer-based systems fall in category of programmable electronic system (PES) as defined in IEC [24]. These systems are used for control, protection or monitoring based on one or more programmable electronic devices [24]. The elements of the system (sensors, processing devices, actuators, power supplies and wiring etc.) are grouped into channels that independently perform(s) a function.

To model the system, most of the assumptions taken for the proposed model are similar as given in Annex B of part-6 of IEC [24]. Assumptions such as (i) failure rate are constant over system life, (ii) channels in a voted group all have the same failure rate, diagnostic rate, diagnostic coverage, Mean time to restore and proof-test interval, are taken unchanged. Some of the assumptions of IEC [24] are modified/generalized. These are given below along with new ones:

1. Overall channel failure rate of a channel is the sum of the dangerous failure rate and safe failure rate for that channel. There values need not to be equal. This is generalization of the assumption [24, 26] that these two failure rates are equal in value.
2. At least one repair team is available to work on all known failure. This is generalization of the assumption in [24, 26]. One repair facility work for one

known failure only. Availability of single repair crew in many cases has been discussed in [30].

3. The fraction of failures specified by diagnostic coverage is detected, corresponding channel is put into a safe state and restored thereafter. This assumption is on contrary to assumption of IEC [24] for low demand case which assumes on line repair, but in case of high demand IEC [24] assumes system achieves safe state after detecting a dangerous fault. With this assumption, *1oo1* and *1oo2* voted group, on any detected fault EUC is put into the safe state.
4. A failure of any kind (*SF, DD, DU*) once occurred to any channel cannot be changed to other types without being restored to healthy state [35]. This means if a channel fails to *SF* state, then unless it is repaired back to healthy state it cannot have failures of type *DD* or *DU* and vice versa.
5. Proof-tests (inspections or functional tests) are conducted on line. Proof-test of a healthy channel neither changes system's state nor EUC's. While a channel with undetected faults is put into safe state following proof-tests. This is a new assumption. It is mainly based on the practice followed in nuclear industry.
6. Proof-tests are periodic with negligible duration. The proof-test interval is at least 3 orders of magnitude greater than diagnostic test interval. This assumption modifies the assumption of IEC [24] which put the limit of 1 order of magnitude. This assumption is based on the fact that order of diagnostic test interval is usually less than or equal to 10s of seconds, while proof-tests interval are not less than a day.
7. Expected interval between demands is at least 3 order of magnitude greater than diagnosis test interval. IEC [24] defines two different limits for low demand and high demand mode of operation. Here limit of high demand operation is taken with limit increased to 3 orders. This is based on the assumption that expected interval between demands is not less than a day [37] even in high demand mode.
8. On occurrence of safe fault, the channel is put into safe state, independent of other channels. Hence all safe failures even in voted groups are detectable.
9. Time between demands is assumed to follow exponential distribution with parameter demand rate. This is as in Bukowski [37].
10. Time to restart the EUC following safety action by safety system on demand is assumed negligible.
11. Following safe failure of safety system, EUC can be started as soon as sufficient number of channels of safety system is operational.
12. The fraction of failures that have a common cause are assumed be equal for both safe and dangerous undetected failures.

State-transition diagram of a generic system is given in Fig. 4.8 .System state *OK* represents healthy state of all its channels. When system has some channels either in *SF* or *DU* state but sufficient number of channels are in healthy state to take safety action on demand is denoted by *Dr*. When more than tolerable number of channels are either in *SF* state or *DU* state, it leads the system to go to F_S or F_{DU} respectively. Demand for safety action when system is in F_{DU} lead to *DEUC* or

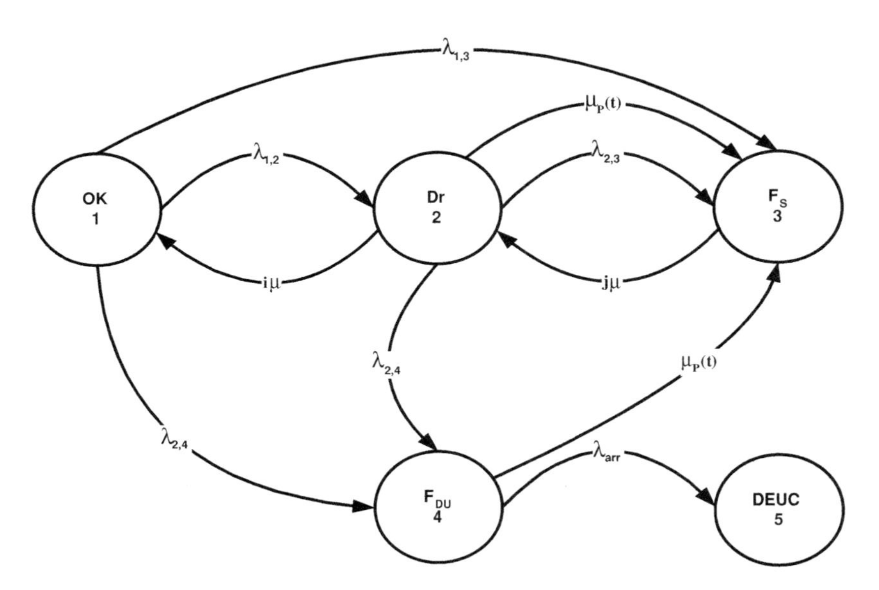

Fig. 4.8 Generic state-transition diagram for a safety system. State *OK* depicts the healthy state, *Dr* degraded working state, F_S safe failure state, F_{DU} unsafe failure state and, *DEUC* damage to EUC state. $\lambda_{i,j}$ denoted the transition from state *i* to *j*, due to failure(s), λ_{arr} is the arrival rate of demands and, $i\mu$ and $j\mu$ is the repair rate from the corresponding states. $\mu_p(t)$ denotes the time-dependent proof-test event

accident condition. Transitions from *OK* state to *Dr* state is represented by $\lambda_{1,2}$. This includes CCF of more than one channel, considering MBF [32]. Failure of all channels due to CCF to safe or dangerous undetected is represented by $\lambda_{1,3}$ and $\lambda_{1,4}$ respectively. Further safe or dangerous faults of healthy channels from system state *Dr*, lead the system to safe sate (F_S) or dangerous undetected state (F_{DU}) respectively. These transition rates are denoted by $\lambda_{2,3}$ and $\lambda_{2,4}$ including CCF more than one healthy channel. By means of repair channels with *SF* get restored to healthy state, this is denoted by $i\mu$. Here i denote either the number of known failures or identical repair facilities whichever is minimum. From F_{DU} system goes to *DEUC* state on demand arrival as represented by λ_{arr}. Proof-test is modeled by $\mu_P(t)$. Proof test convert the channels with *DU* failure to go to *SF*.

The state-transition diagram of Fig. 4.8 explicitly considers safe failures. Refs. [25, 28, 31] also model safe failure state for different purposes. Here the intention is to be able to estimate probability of being in all possible states up to a specified time. In addition to IEC [24, 26, 27], periodic inspections (proof-tests) are considered in [25]. Bukowski [25] defined $MTTF_D$ & $MTTF_S$. While estimating $MTTF_S$ (or $MTTF_D$) it assumes repair (or restoration) from $MTTF_D$ (or $MTTF_S$). Means, it give mean time to reach safe (or danger) state irrespective of the number of visits to dangerous failure (or safe) state. Similarly it defines availability as *probability system is successfully operating at time t without regard to previous failure or repair.*

Demand rate is incorporated in [37] as in the proposed model. If system is in dangerous failure state on demand arrival, then system goes to a state similar to *DEUC* of our model. Except modeling of demand, model of ref. [37] is totally different than the one considered here. The key differences are as follows [39–41]:

1. change of safety system state on demand arrival in healthy states
2. considers online repair
3. does not incorporate safe failures
4. does not incorporate CCF
5. does not incorporated periodic proof-tests (or inspection or functional test)

The proposed model can be considered combining, in conceptual sense, the ideas of periodic proof-test [25] and process demand rate [37].

4.5.4 Performance Based Safety and Availability Indices

All transitions in Markov model of Fig. 4.8, except $\mu_P(t)$ are constant and independent of time. Exclusion of this transition (i.e. $\mu_P(t) = 0$ for all t) from state-transition diagram, transforms the state transition diagram in to a continuous-time Markov chain (CTMC) [7, 42]. The infinitesimal generator matrix, Q of the CTMC of Fig. 4.8 is given by,

$$Q = \begin{bmatrix} \Lambda_{TT} & 0 \\ \Lambda_{TA} & 0 \end{bmatrix} \tag{4.9}$$

The CTMC of Fig. 4.8 is absorbing, so its infinitesimal matrix, Q, is singular, i.e $|Q| = 0$. To analyze such a system analytically, use of Darroch and Seneta [43] technique of only considering transient state is proposed. Λ_{TT} is the infinitesimal generator matrix of CTMC considering transient states only. All transient state of the CTMC communicate to absorbing state, ensures Λ_{TT} is regular (i.e. $\Lambda_{TT} \neq 0$) [43]. Time dependent transient state probabilities are given by solving following Chapman–Kolmogrov equation [7, 42].

$$\begin{aligned} \dot{\pi}(t) &= \Lambda_{TT}\pi(t) \\ \pi(t) &= [P_1(t)\ P_2(t)\ P_3(t)\ P_4(t)] \end{aligned} \tag{4.10}$$

Solution of (4.10) gives time varying transient state probabilities without periodic proof-tests. Incorporation of periodic proof-tests in this model makes it a non-Markovian model. Marsan [44] proposed a method to analyze a non-Markovian system for steady state which satisfies following two conditions:

1. non-Markovian event is deterministic
2. only one deterministic event is enabled at any instant

Varsha [45] uses a method based on Markov regenerative process [42] to solve availability problems with periodic load pattern. As per Markov regenerative process, time instances, in present context, T_{proof}, $2T_{\text{proof}}$, $3T_{\text{proof}}\ldots$, are called Markov regeneration epochs. State probabilities of the model can be obtained by sequentially solving the Markov chain between Markov regenerative epochs and redistributing the state probabilities at regeneration epochs. Bukowski [25] also uses the similar method to determine MTTF under various conditions, and call it piece-wise CTMC method. Here, Markov regenerative process based analysis to determine state probabilities with periodic proof-tests has been employed.

State probabilities for time up to first regenerative epoch can be obtained from (4.10).

$$\pi(t) = e^{\Lambda_{TT} t}\pi(0) \quad 0 \le t < T_{\text{proof}} \tag{4.11}$$

State probability just first before proof-test are given by

$$\pi(\tau_{-}) = e^{\Lambda_{TT}\tau}\pi(0) \quad \tau = T_{\text{proof}} \tag{4.12}$$

Let state probability redistribution matrix is given by

$$\Delta = [\delta_{ij}] \tag{4.13}$$

Redistribution matrix (Δ) is a square matrix. The rows of this matrix correspond to each state of the Markov model. In any state, channels of the system can have OK, SF or DU states. Values of elements of redistribution matrix (Δ_{ij}) are determined as follows:

$$\delta_{ij} = \begin{cases} 1 & \forall i = j \text{ and all channels are either OK of SF in system state corresponding to i} \\ (1-\eta) & \forall i = j \text{ and any channel state is DU in system state corresponding to i} \\ \eta & \forall \text{ state i is having channels with SF, state j having channels with DU,} \\ & \text{and conversion of all channels with DU of state j to SF leads to state i} \\ 0 & \text{otherwise} \end{cases} \tag{4.14}$$

where η is measure of the degree of perfection of proof-tests.

State probabilities just following the first proof test is given by

$$\pi(\tau_{+}) = \Delta e^{\Lambda_{TT}\tau}\pi(0) \quad \tau = T_{\text{proof}} \tag{4.15}$$

and

$$\pi(t) = \pi(\tau + S) = e^{\Lambda_{TT}S}\Delta e^{\Lambda_{TT}\tau}\pi(0) \quad \tau < t < 2\tau; \quad \tau = T_{\text{proof}} \tag{4.16}$$

Generalization of the above equation gives,

$$\pi(t) = \pi(n\tau + S) = e^{\Lambda_{TT}S}\alpha^{n}\pi(0) \quad n\tau < t < (n+1)\tau \tag{4.17}$$

where $\alpha = \Delta e^{\Lambda_{TT}\tau}$.

Let system operates continuously for time duration of T, then mean state probabilities for this duration can be computed as:

$$E[\pi(t)] = \bar{\pi} = \frac{\int_0^T \pi(t)\mathrm{d}t}{\int_0^T \mathrm{d}t} \tag{4.18}$$

$$\bar{\pi} = \frac{1}{T}\left(\sum_{j=1}^{n} \int_{(j-1)\tau}^{j\tau} \pi(t)\mathrm{d}t + \int_{n\tau}^{n\tau+S} \pi(t)\mathrm{d}t\right) \tag{4.19}$$

$$\bar{\pi} = \frac{1}{T}\left[\Lambda_{TT}^{-1}\left([e^{\Lambda_{TT}\tau} - I][I-\alpha]^{-1}[I-\alpha^n] + [e^{\Lambda_{TT}S} - I]\alpha^n\right)\pi(0)\right] \tag{4.20}$$

where $T = n\tau + s$.

Equation 4.20 gives the closed form solution for state probabilities considering demand rate and periodic proof-test (perfect as well as imperfect).

4.5.4.1 Safety Index: Probability of Failure on Actual Demand (*PFaD*)

The safety index probability of failure on actual demand (*PFaD*) is intended to measure the probability of reaching *DEUC* state. $\pi(t)$ of (4.17) gives the probability of all the states except *DEUC*. From the state-transition diagram of Fig. 4.8, it is clear that system is conservative. Sum of all state probabilities at any instant shall be '1', i.e. system shall be in either of 5 states. The probability of not being in states, defined in $\pi(t)$, is *DEUC* state probability, i.e. *PFaD*. $PFaD(t)$ is given by,

$$PFaD(t) = 1 - [1_x]\pi(t) \tag{4.21}$$

where 1_x is the vector of 1s, equal to size of $\pi(t)^{\mathrm{T}}$.

Average probability of failure on demand, *PFaD* can be evaluated using (4.20) and (4.21)

$$mean\, PFaD = 1 - [1_x]\bar{\pi} \tag{4.22}$$

where 1_x is the vector of 1s equal to size of $\bar{\pi}^{\mathrm{T}}$.

The Markov model being absorbing $\pi(t)$ will decrease to 0 with increasing time. So, $PFaD(t)$ like failure distribution (complement of reliability) [7] is a non-decreasing function of time.

4.5.4.2 Comparison with *PFDPRS*

As discussed earlier, formulation of *PFaD* can be considered as combination of research work of [25] and [37]. Ref. [25] uses index (*MTTF*) while [37] uses a similar performance based safety index, $PFDPRS(t)$. $PFDPRS(t)$ does not include periodic proof-test. When proof-tests are not considered in safety index *PFaD(t)* then $PFaD(t)$ and $PFDPRS(t)$, by definition, are same except differences in respective models.

Table 4.1 Parameter values used for comparison of *PFaD(t)* and *PFDPRS(t)* [37]

Parameter	Value
λ_S	$0\,\text{h}^{-1}$
λ_D	$2 \times 10^{-6}\,\text{h}^{-1}$
DC	0.9
η	1
μ	$1/8\,\text{h}^{-1}$
T	5,000 h
T_{Proof}	6,000 h

Table 4.2 Comparison of results, *PFaD(t)* and *PFDPRS(t)* [37]. First column shows MTBD (mean time between demands), second column shows value of safety index *PFDPRS(t)*, PFaD(t) and its mean value are given in third and fourth column, % relative difference between PFDPRS(t) and PFaD(t) are given in column 5

MTBD	PFDPRS(t) [37]	PFaD(t)	PFaD	% difference
1/day	0.0015	0.000995	0.000495	33.688667
1/week	0.0014	0.000966	0.000467	31.007143
1/month	0.00085	0.000856	0.000377	−0.6752941
1/year	0.00022	0.000238	0.000083	−0.3636364
1/10 years	0.00002	0.000028	0.000009	−0.39965

To compare $PFaD(t)$ values with that of $PFDPRS(t)$ [37], parameter values of [26] are taken. Ref. [26] does not consider safe failure hazard rate and periodic proof-tests. Safe failures hazard rate are taken to be zero and proof-test interval is taken more than operation time for analysis. For convenience all parameter values are given in Table 4.1.

Result of the two models for different demand rates are given in Table 4.2. At higher demand rate, $PFaD(t)$ values are less than that of corresponding $PFDPRS(t)$ values. The main reason for this is difference is Bukowski's [37] model considers online repair and unsafe failure from dangerous detected state. At low demand rate, both models are in good agreement.

Mean *PFaD* values for the specified time duration, $T = 5{,}000$ h is also given in Table 4.2. For the parameter values of Table 4.1, avg. *PFD* and *PFH* values are 5.25×10^{-3} and 2×10^{-7}, respectively. These two values are the probability of being in dangerous undetected state in low demand mode and high demand mode. While the safety index avg. *PFaD* for the specified operational time gives the actual probability of failing on demand or probability of damage to EUC.

4.5.4.3 Availability Index: Manifested Availability

$PFaD(t)$ denotes the fraction of system failed on demand by time t. Failing on demand often lead to endangering plant or EUC. Safety systems which failed on demand can not be brought back to operation.

Let a safety system survives from unsafe failures (i.e. *DEUC*) up to time t, then conditional state probability are given as:

$$\bar{\pi}_i(t) = \frac{\pi_i(t)}{\sum_i \pi_i(t)} = \frac{\pi_i(t)}{1 - PFaD(t)} \tag{4.23}$$

With this condition, probability of not being in FS gives the availability of safety system. This availability is termed as *manifested Availability* (*mAv*). Average *mAv* value up to time $'t'$ is given as:

$$avg.mAv(t) = 1 - \sum_i \hat{\pi}_i(t) \quad \text{where system state corresponding to } i \in F_S \tag{4.24}$$

This definition of availability is different from Bukowski [25], which considers probability of being in state *OK* (with reference to Markov model of Fig. 4.8), is availability. *mAv* takes into account both types of failures affecting availability of EUC.

4.6 Examples

4.6.1 Example 1

4.6.1.1 System Description

To illustrate effect of proof-test interval along with demand rate, a hypothetical protection system of a nuclear reactor is taken. The purpose of the system is to take safety action (shutdown the reactor) at demand. Composition and number of modules required to configure one channel is shown in Fig. 4.9. Channel uses in house developed programmable electronic modules. Module 8687EURO is processor module containing x86 processor and math co-processor. SMM256 contains EPROMs and RAMs. EPROMs are used for storing program while RAM is used as scratch pad area. Modules RORB, DIFIT, DOSC and ADA12 are input/output modules. These are used for acquiring inputs and generating outputs. Module WDT (watchdog timer) is used to ensure a channel's outputs go to safe failure

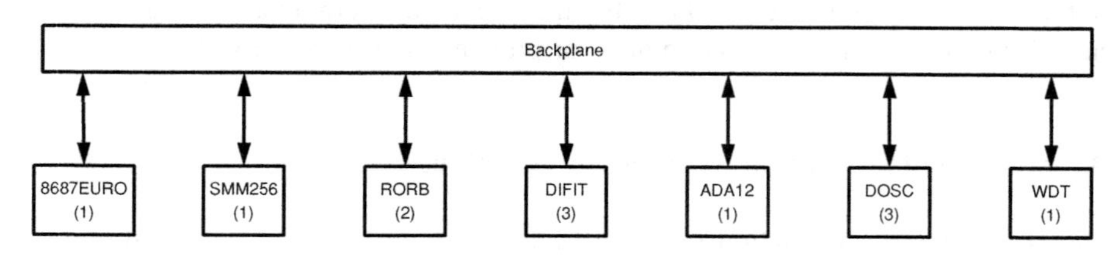

Fig. 4.9 Composition of a channel. Number within braces shows the quantity of such modules in a channel

Table 4.3 Module hazard rate and calculation of channel hazard rate

S. No.	Module name	Quantity	Module hazard rate	Total hazard rate
1	8687EURO	1	1.22E-05	1.22E-05
2	SMM-256	1	4.42E-06	4.42E-06
3	DIFIT	3	1.07E-05	3.21E-05
4	RORB	2	3.03E-06	6.05E-06
5	ADA-12	1	6.73E-06	6.73E-06
6	DOSC	3	7.11E-06	2.13E-05
7	WDT	1	2.80E-06	2.80E-06
8	Backplane	1	0	0
		Channel hazard rate (λ)		8.56E-05

Table 4.4 Diagnostic parameters

S. No.	Parameter	Value
1	Fraction dangerous	0.48
2	Diagnostic coverage (DC)	0.74
3	Demand rate	5.0×10^{-4} h^{-1}
4	Diagnosis rate	0.1 h^{-1}
5	Repair rate	1 h^{-1}
6	Proof-test interval	2,000 h

state, when a dangerous failure is detected in the channel. Module failure (hazard) rates are taken from Khobare et al. [46]. Based on module hazard rates channel hazard rate is estimated assuming any module failure lead to channel failure.

Protection system is configured as *2oo3* i.e. TMR (Triple Modular Redundant). The *2oo3* system configuration is shown in Fig. 4.12b. Three channels operate independently and open their control switches to shutdown the EUC. Control switches from individual channels are wired to form a *2oo3* majority voting logic. This enables the system to tolerate one channel's failure of either type safe or unsafe.

Ratio of safe failure to unsafe failures and coverage factor are taken from [47]. Module hazard rate values are given in Table 4.3, diagnostic parameters given in Table 4.4 and derived parameters required for model are given in Table 4.5.

4.6.1.2 Model

The first step is derivation of Markov model for the system. Markov model of the system is shown in Fig. 4.10. States are marked with 3-tuple, (i, j, k), i shows the number of healthy channels, j channel(s) in safe failure state and k channel(s) in dangerous failure state. Transition rate is in terms of safe, dangerous and repair rate. MATLAB code for the example is given below:

```
% ***********Input Parameters *******************
prmtr = load('parameters2oo3w.txt');

LSafe = prmtr(1,1);
LDang = prmtr(2,1);
MeanRepairTime = prmtr(3,1);

DiagCov = prmtr(4,1);
PrTestCov = prmtr(5,1);

Tproof = prmtr(6,1);
RunTime = prmtr(7,1);
MeanTimeBetweenDemands = prmtr(8,1);

CommonCause2 = prmtr(9,1);
CommonCause3 = prmtr(10,1);

% **********Derived Parameters ******************
L1 = LSafe + DiagCov*LDang;
L2 = (1-DiagCov)*LDang;
M = 1/MeanRepairTime;
La = 1/MeanTimeBetweenDemands;
a = PrTestCov;
B = CommonCause2;
B2 = CommonCause3;
Tp = Tproof;
n = floor(RunTime/Tp);
s = RunTime - n*Tp;

B2_ = 1 - B2;
B_ = 1 -B;
Alfa = 1-(2-B2)*B;
L = L1 + L2;

% Definition of infinitesimal generator matrix for 1oo1
Q = [ -(3*Alfa+3*B2_*B+B2*B)*L       M              0        0
       0       0        0        0       0        0;
       3*Alfa*L1                -(2*B_+B)*L-M       M        0
             0       0        0       0        0        0;
       3*B2_*B*L1                     2*B_*L1        -(L+M)       M
             0       0        0       0        0        0;
       B2*B*L1                        B*L1               L1      -M
             0       0        0       0        0        0;
       3*Alfa*L2                      0                  0        0
         -(2*B_+B)*L   M        0        0       0        0;
       0                              2*B_*L2            0        0
           2*B_*L1   -(L+M)     M        0       0        0;
       0                              0                  L2       0
             B*L1      L1       -M        0       0        0;
       3*B2_*B*L2                     0                  0        0
             2*B_*L2   0        0     -(L+La)    M        0;
       0                              B*L2               0        0
             0         L2       0        L1    -(M+La)    0;
       B2*B*L2                        0                  0        0
             B*L2      0        0        L2       0      -La;];

% Definition of Delta matrix
Delta = [1 0 0 0    0    0    0    0    0    0;
         0 1 0 0    a    0    0    0    0    0;
         0 0 1 0    0    a    0    a    0    0;
         0 0 0 1    0    0    a    0    a    a;
         0 0 0 0  1-a    0    0    0    0    0;
         0 0 0 0    0 (1-a)   0    0    0    0;
         0 0 0 0    0    0 (1-a)   0    0    0;
         0 0 0 0    0    0    0 (1-a)   0    0;
         0 0 0 0    0    0    0    0 (1-a)   0;
         0 0 0 0    0    0    0    0    0 (1-a);];

P0 = [1 0 0 0 0 0 0 0 0 0]';
```

```
% ***************************************************
E = Delta*expm(Q*Tp);

I = eye(size(Q));

% EPn = (1/RunTime)*(inv(Q)*(expm(Q*s)-I)*inv(I-E)*(I-E^(n+1))*P0);

Temp = inv(Q)*(expm(Q*Tp)-I)*inv(I-E)*(I-E^n)*P0;

meanProb =(1/RunTime)*(Temp + inv(Q)*(expm(Q*s)-I)*P0);

anaPFaD =1 - ones(size(P0'))*(meanProb);

anaFs = [0 0 1 1 0 0 1 0 0 0]*meanProb;

anamAv = 1 - (anaPFaD + anaFs);


for i = 0:RunTime/10,
    Time(i+1) = 10*i;
    n = floor(Time(i+1)/Tp);
    s = Time(i+1)-n*Tp;
    Pn(i+1) = (ones(size(P0')))*expm(Q*s)*E^n*P0;
    F(i+1) = 1-Pn(i+1);
end

% PFaD_t = 1 - ones(size(Pn'))*Pn;
% PFaD = 1 - ones(size(EPn'))*EPn;
% [Tp PFaD]

plot(Time, F);

[anaPFaD anaFs anamAv]
Ratio = F(size(F,2))/anaPFaD
%[PFaD_t PFaD]
```

Table 4.5 Derived parameter values for safety model

Parameter	Value
λ_S	4.45×10^{-5} h^{-1}
λ_D	4.10×10^{-5} h^{-1}
DC	0.74
η	1
μ	1 h^{-1}
T_{proof}	[100, 100, 10,000] h
λ_{arr}	[2/year, 1/year, 1/5year, 1/10year]
T	6 year

4.6.1.3 Results

Mean *PFaD* values for different demand rates and proof-test interval are plotted in Fig. 4.11. If required value of *PFaD* is assumed 1×10^{-4}. From the figure, it can be observed to meet this target safety value frequent proof-test are required. At design time demand rate of a new EUC is not known clearly. So this plot can be used to choose a proof-test interval which guarantees the required *PFaD* value for maximum anticipated demand.

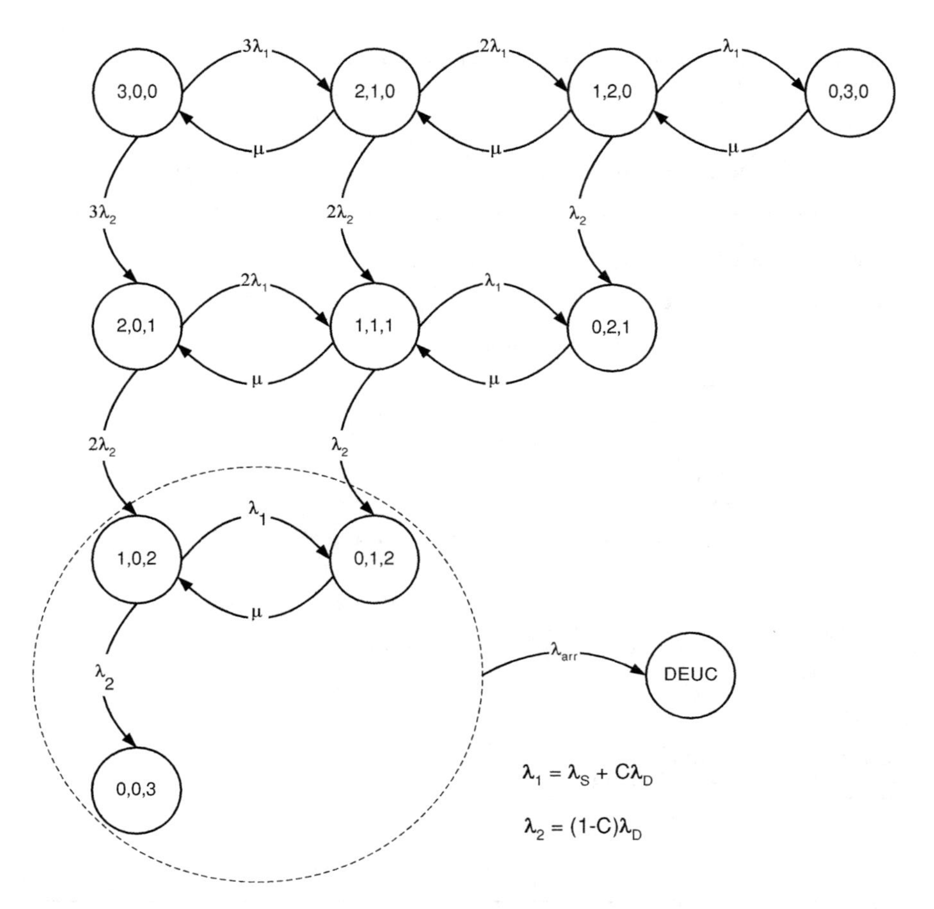

Fig. 4.10 Markov model for safety analysis of Example 1. This safety model is based on generic safety model shown in Fig. 4.8. State {(3,0,0)} corresponds to *OK* state, states {(2,1,0), (2,0,1), (1,1,0)} corresponds to *Dr* state and, states {(1,2,0), (0,3,0), (0,2,1)} corresponds to F_S state of Fig. 4.8. States {(1,0,2), (0,1,2), (0,0,3)} corresponds to F_{DU} states. System may fail dangerously from any of the F_{DU} states. Transition rate λ_{arr} from each of these state is same due to distribution property of transitions, i.e. $(P_{102} + P_{012} + P_{003}).\lambda_{arr} = P_{102}.\lambda_{arr} + P_{012}.\lambda_{arr} + P_{003}.\lambda_{arr}$

4.6.2 Example 2

To illustrate evaluation of *PFaD* and manifested availability, 2 commonly used hardware architectures; *1oo2* and *2oo3* of a safety system are chosen. Schematic of these architectures are shown in Fig. 4.12.

1oo2 architecture consists of 2 *s*-identical channels with output switches wired in series, also called output ORing. If either channel is *SF*, then its control switch opens and EUC is shutdown. Therefore, *1oo2* architecture is sensitive to safe failures of its channels: even a single channel's safe failure causes the EUC to shutdown. On the other hand, if one channel is in F_{DU}, then its switch is unable to

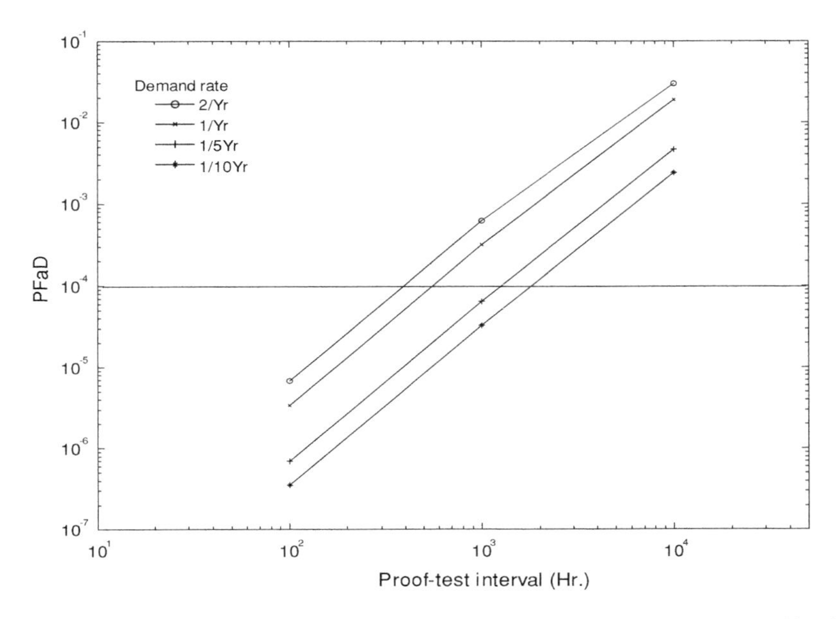

Fig. 4.11 *PFaD* values with respect to proof-test interval at different demand rates. This plots relationship between mean *PFaD* and proof-test interval for 4 different demand rates

open at demand, but other channel's switch open and EUC shutdown is ensured. So, *1oo2* system can endanger EUC, only if both channels are in *DU* at demand. So, *1oo2* with output ORing can tolerate one channel's *DU* failures only. Both channels can fail to *SF* or *DU* due to CCF. Beta-factor model [24, 32] for CCF is used. The Markov model of the system with one repair station is shown in Fig. 4.13.

The states of the Markov chain are denoted by a 3-tuple, (i, j, k)

- i denotes the number of channels in *OK* state
- j denotes the number of channels in *SF* state
- k denote the number of channels on *DU* state

2oo3 consists of 3 *s*-identical channels with a pair of output control switches from each channel. These control switches are used in implementing majority voting logic. *2oo3* architecture shutdowns the EUC when two channels go to *SF*, It endangers the EUC when 2 channels are in dangerous failure states (*DU*) at demand. Means, *2oo3* can tolerate one channel's safe or dangerous failure. To incorporate CCF we have used MBF (multiple beta factor) of Hokstad [32]. The advantage of MBF is it can model variety of cases, beta-factor, gamma-factor, and base case. Beta-factor allows only simultaneous failure of three channels, gamma-factor allows simultaneous failure of two channels only, while base case allows combination of simultaneous failure of two and three channels. The Markov model of the system with one repair station is shown in Fig. 4.14.

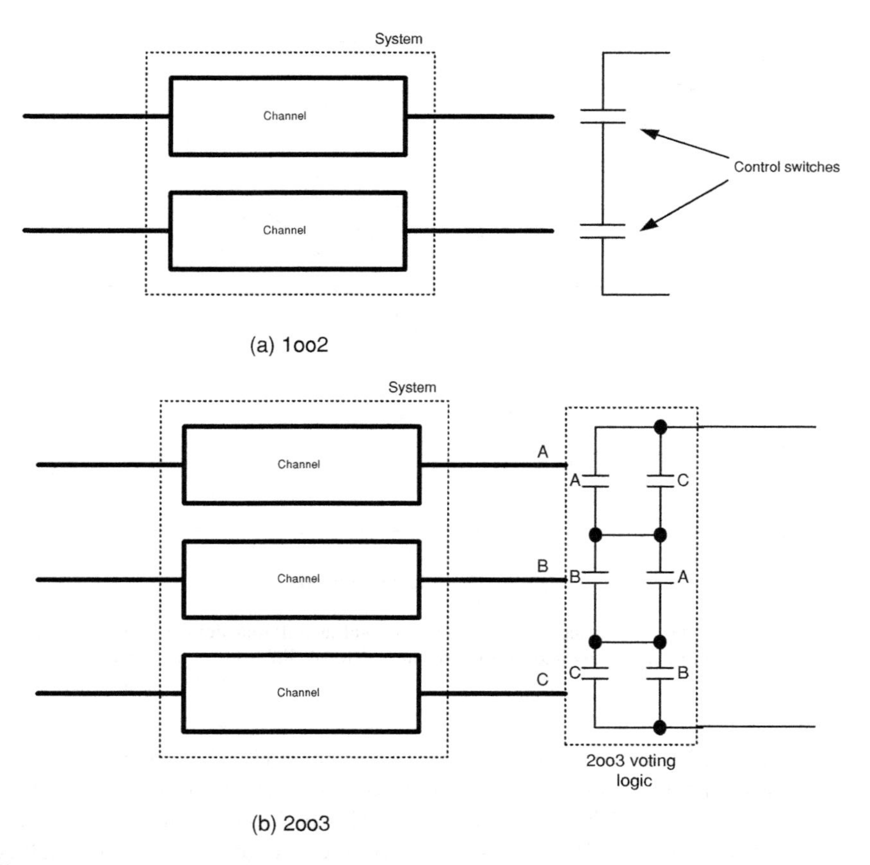

Fig. 4.12 Schematic of *1oo2* and *2oo3* architectures

4.6.2.1 Parameter Values

Table 4.6 gives system parameter values such as the channel hazard rate, repair along with proof-test interval or mission time used for the example architectures.

Probability redistribution matrix, Δ, for 2 architectures is derived from discussion of Sect. 4.5.3. These are given as below:

For *1oo2*

$$\Delta = \begin{bmatrix} 1 & 0 & 0 & 0 & 0 & 0 \\ 0 & 1 & 0 & 0.9 & 0 & 0 \\ 0 & 0 & 1 & 0 & 0.9 & 0.9 \\ 0 & 0 & 0 & 0.1 & 0 & 0 \\ 0 & 0 & 0 & 0 & 0.1 & 0 \\ 0 & 0 & 0 & 0 & 0 & 0.1 \end{bmatrix} \tag{4.25}$$

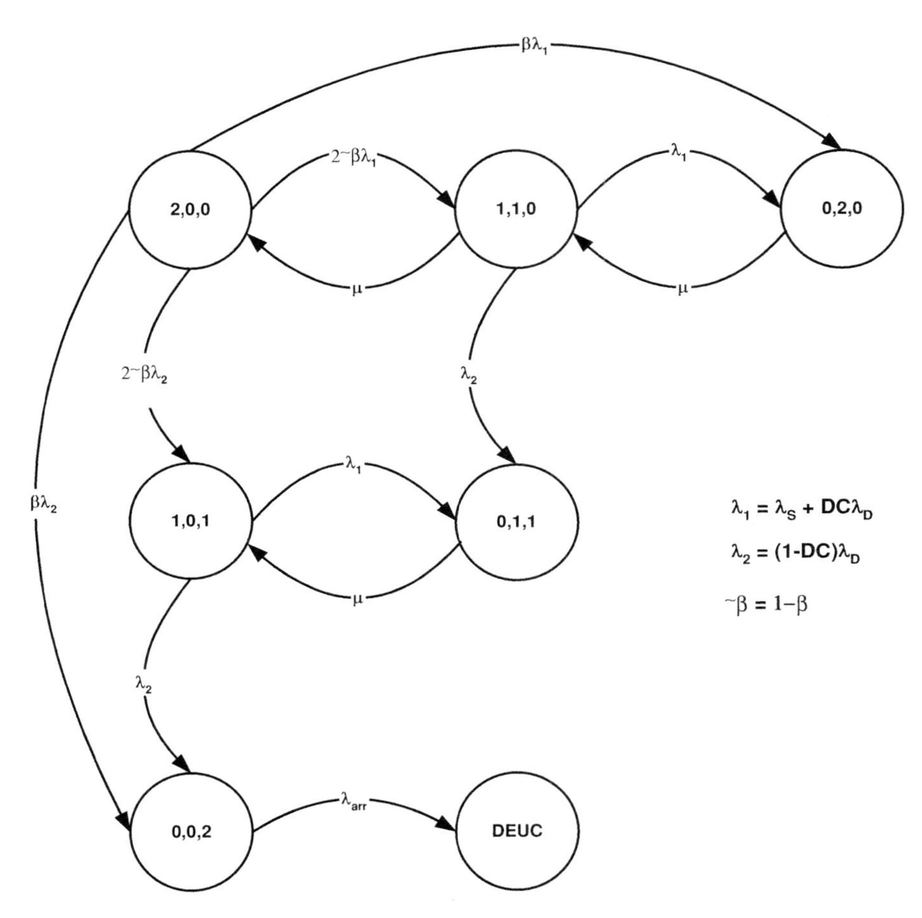

Fig. 4.13 Markov model of *1oo2* system of Example 2. Markov model is based on generic Markov model of Fig. 4.8. State {(2,0,0)} corresponds to *OK*, {(1,0,1)} to *Dr*, {(1,1,0), (0,2,0), (0,1,1)} to F_S and {(0,0,2)} to *DU* state of Fig. 4.8. β represents the fraction of failures due to CCF. λ_1 is hazard rate of safe failures, while λ_2 is hazard rate of dangerous failures

For *2oo3*

$$\Delta = \begin{bmatrix} 1 & 0 & 0 & 0 & 0 & 0 & 0 & 0 & 0 & 0 \\ 0 & 1 & 0 & 0 & 0.9 & 0 & 0 & 0 & 0 & 0 \\ 0 & 0 & 1 & 0 & 0 & 0.9 & 0 & 0.9 & 0 & 0 \\ 0 & 0 & 0 & 1 & 0 & 0 & 0.9 & 0 & 0.9 & 0.9 \\ 0 & 0 & 0 & 0 & 0.1 & 0 & 0 & 0 & 0 & 0 \\ 0 & 0 & 0 & 0 & 0 & 0.1 & 0 & 0 & 0 & 0 \\ 0 & 0 & 0 & 0 & 0 & 0 & 0.1 & 0 & 0 & 0 \\ 0 & 0 & 0 & 0 & 0 & 0 & 0 & 0.1 & 0 & 0 \\ 0 & 0 & 0 & 0 & 0 & 0 & 0 & 0 & 0.1 & 0 \\ 0 & 0 & 0 & 0 & 0 & 0 & 0 & 0 & 0 & 0.1 \end{bmatrix} \quad (4.26)$$

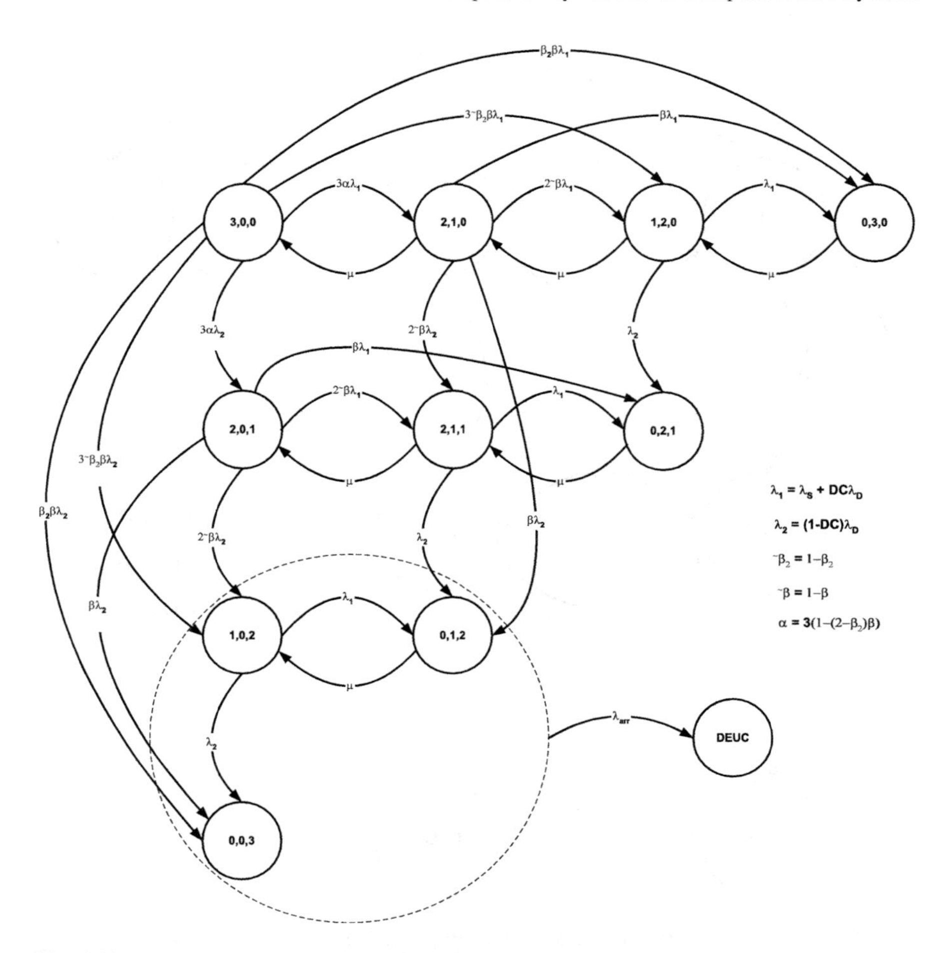

Fig. 4.14 Markov model of *2oo3* system of Example 2. Markov model is based on generic Markov model of Fig. 4.8. State {(3,0,0)} corresponds to *OK*, {(2,1,0), (2,0,1), (2,1,0)} to *Dr*, {(1,2,0), (0,3,0), (0,2,1)} to F_s and {(1,0,2), (0,1,2), (0,0,3)} to *DU* state of Fig. 4.8. β and β_2 represents the fraction of failures due to CCF, α is fraction of individual failures. λ_1 is hazard rate of safe failures, while λ_2 is hazard rate of dangerous failures

MATLAB code for the example is given below:

```
% ***********Input Parameters ********************
prmtr = load('parameters1oo2O.txt');

LSafe = prmtr(1,1);
LDang = prmtr(2,1);
MeanRepairTime = prmtr(3,1);

DiagCov = prmtr(4,1);
PrTestCov = prmtr(5,1);

Tproof = prmtr(6,1);
```

```
RunTime = prmtr(7,1);
MeanTimeBetweenDemands = prmtr(8,1);

CommonCause2 = prmtr(9,1);
CommonCause3 = prmtr(10,1);
NumberofRepairStn = prmtr(11,1);

% **********Derived Parameters *******************
L1 = LSafe + DiagCov*LDang;
L2 = (1-DiagCov)*LDang;
M = 1/MeanRepairTime;
La = 1/MeanTimeBetweenDemands;
a = PrTestCov;
B = CommonCause2;
B2 = CommonCause3;
Tp = Tproof;
n = floor(RunTime/Tp);
s = RunTime - n*Tp;
N = NumberofRepairStn;

B2_ = 1 - B2;
B_ = 1 -B;
Alfa = 1-(2-B2)*B;
L = L1 + L2;
% Definition of infinitesimal generator matrix for 1oo1
Q = [   -(2*B_+B)*L       M        0        0        0        0;
          2*B_*L1      -(L+M)      M        0        0        0;
          B*L1            L1      -M        0        0        0;
          2*B_*L2         0        0       -L        M        0;
          0               L2       0        L1      -M        0;
          B*L2            0        0        L2       0      -La;];

% Definition of Delta matrix
Delta = [1 0 0   0    0   0;
         0 1 0   a    0   0;
         0 0 1   0    a   a;
         0 0 0 (1-a)   0   0;
         0 0 0   0  (1-a)  0;
         0 0 0   0    0  (1-a);];


P0 = [1 0 0 0 0 0]';

% ************************************************
E = Delta*expm(Q*Tp);

I = eye(size(Q));

%EPn=(1/RunTime)*(inv(Q)*(expm(Q*s)-I)*inv(I-E)*(I-E^(n+1))*P0);

Temp = inv(Q)*(expm(Q*Tp)-I)*inv(I-E)*(I-E^n)*P0;

meanProb =(1/RunTime)*(Temp + inv(Q)*(expm(Q*s)-I)*P0);
anaPFaD =1 - ones(size(P0'))*(meanProb);

anaFs = [0 1 1 0 1 0]*meanProb;

anaS = [1 0 0 1 0 0]*meanProb;


for i = 0:RunTime/10,
    Time(i+1) = 10*i;
    n = floor(Time(i+1)/Tp);
    s = Time(i+1)-n*Tp;
```

```
    Pn(i+1) = (ones(size(P0')))*expm(Q*s)*E^n*P0;
    F(i+1) = 1-Pn(i+1);
end

% PFaD_t = 1 - ones(size(Pn'))*Pn;
% PFaD = 1 - ones(size(EPn'))*EPn;
% [Tp PFaD]

plot(Time, F);

[anaPFaD anaFs anaS]

Ratio = F(size(F,2))/anaPFaD
%[PFaD_t PFaD]
```

Table 4.6 Parameter values for Example 2

Parameter	Value
λ_S	5×10^{-6} h^{-1}
λ_D	5×10^{-6} h^{-1}
DC	0.9
η	0.9
μ	1/8 h^{-1}
β	0.1
β_2	0.3
T	87,600 h
MTBD	43,800 h

4.6.2.2 Calculation and Results

Average *PFaD* and availability values for the specified period are evaluated. These values are evaluated with parameter values given in Table 4.6 for proof-test interval [100 h, 12,000 h] at increment of 100 h. The plots of average *PFaD* and *mAv* for *1oo2* and *2oo3* architectures are given in Figs. 4.15 and 4.16 respectively. Step-wise detailed calculation is given below for better clarity:

Case-1 1oo2 architecture.

Step 1 First step is to determine infinitesimal generator matrix, Q, which is composed of Λ_{TT} and Λ_{TA}. Λ_{TT} is infinitesimal generator matrix for transient states, {(2,0,0), (1,1,0), (0,2,0), (1,0,1), (0,1,1), (0,0,2)}, and Λ_{TA} is infinitesimal generator matrix for absorbing state DEUC.

$$Q = \begin{bmatrix} \Lambda_{TT} & 0 \\ \Lambda_{TA} & 0 \end{bmatrix}$$

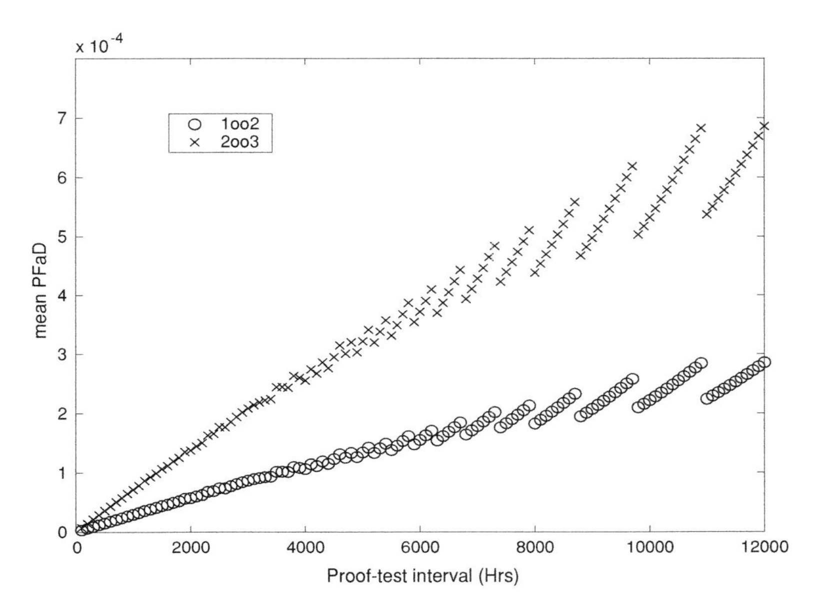

Fig. 4.15 Variation of avg. *PFaD* w.r.t. T_{proof}. For both hardware architectures mean *PFaD* is evaluated for proof-test interval [100 h, 12,000 h] at a increment of 100 h for operating time of $T = 87{,}600$ h. So, each point gives mean *PFaD* value for corresponding proof-test τ (T_{proof}) value. This saw tooth behavior is observed because of 'S' of (4.20). With varying proof-test interval, 'S' may have value from 0 to approx. equal to proof-test interval. For proof-test values, at which 'S' is zero, mean PFaD be local minimum

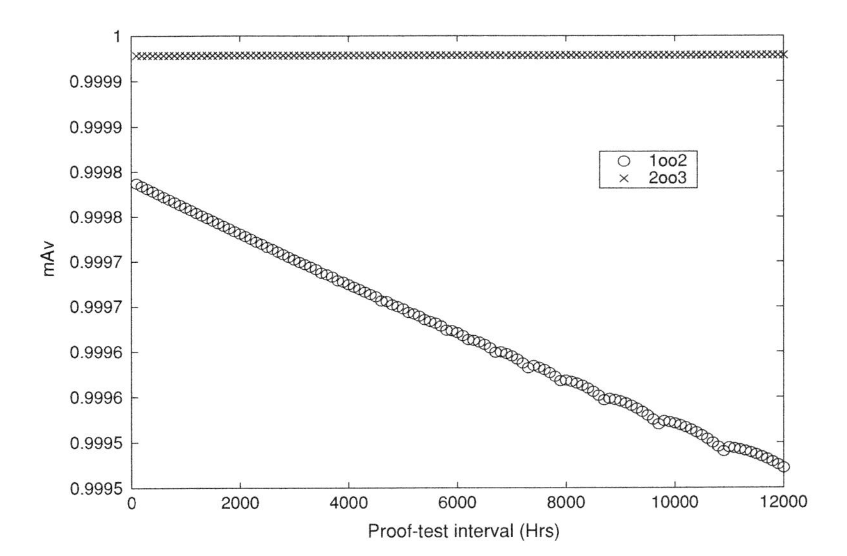

Fig. 4.16 Variation of avg. mAv w.r.t. T_{proof}

From the Fig. 4.13, λ_{TT} and λ_{TA} is given as:

$$\Lambda_{TT} = \begin{matrix} P_{200} \\ P_{110} \\ P_{020} \\ P_{101} \\ P_{011} \\ P_{002} \end{matrix} \begin{bmatrix} -((\beta+2\tilde{\beta})\lambda_1+(\beta+2\tilde{\beta})\lambda_2) & \mu & 0 & 0 & 0 & 0 \\ 2\tilde{\beta}\lambda_1 & -(\lambda_1+\lambda_2+\mu) & \mu & 0 & 0 & 0 \\ \beta\lambda_1 & \lambda_1 & -\mu & 0 & 0 & 0 \\ 2\tilde{\beta}\lambda_2 & 0 & 0 & -(\lambda_1+\lambda_2) & \mu & 0 \\ 0 & \lambda_2 & 0 & \lambda_1 & -\mu & 0 \\ \beta\lambda_2 & 0 & 0 & \lambda_2 & 0 & -\lambda_{arr} \end{bmatrix}$$

$$\Lambda_{TA} = \begin{matrix} P_{200} \\ P_{110} \\ P_{020} \\ P_{101} \\ P_{011} \\ P_{002} \end{matrix} \begin{bmatrix} 0 \\ 0 \\ 0 \\ 0 \\ 0 \\ \lambda_{arr} \end{bmatrix}$$

From the parameter values given in Table 4.6, Λ_{TT} and Λ_{TA} is given as:

$$\Lambda_{TT} = \begin{bmatrix} -1.9\times10^{-5} & 1.25\times10^{-1} & 0 & 0 & 0 & 0 \\ 1.71\times10^{-5} & -1.25\times10^{-1} & 1.25\times10^{-1} & 0 & 0 & 0 \\ 9.5\times10^{-7} & 9.5\times10^{-6} & -1.25\times10^{-1} & 0 & 0 & 0 \\ 9.0\times10^{-7} & 0 & 0 & -1.0\times10^{-5} & 1.25\times10^{-1} & 0 \\ 0 & 5.0\times10^{-7} & 0 & 9.5\times10^{-6} & -1.25\times10^{-1} & 0 \\ 5.0\times10^{-8} & 0 & 0 & 5.0\times10^{-7} & 0 & -2.283\times10^{-5} \end{bmatrix}$$

Similarly

$$\Lambda_{TA} = \begin{bmatrix} 0 \\ 0 \\ 0 \\ 0 \\ 0 \\ 2.283\times10^{-5} \end{bmatrix}$$

Step 2 Evolution of transient states with time is given by:

$$\dot{\pi}(t) = \Lambda_{TT}\pi(t)$$

$$\pi(t) = \begin{bmatrix} P_{200}(t) \\ P_{110}(t) \\ P_{020}(t) \\ P_{101}(t) \\ P_{011}(t) \\ P_{002}(t) \end{bmatrix}$$

and

$$\pi(t) = \begin{bmatrix} 1 \\ 0 \\ 0 \\ 0 \\ 0 \\ 0 \end{bmatrix}$$

Step 3 Let proof-test interval, T_{proof} is 100 h, then state-probabilities just before Ist proof-test, for example 99 h, is given by:

$$\pi(\tau_{-}) = e^{\Lambda_{TT}\tau}\pi(0) \quad \tau = T_{\text{proof}}$$

$$\pi(99) = \begin{bmatrix} 9.99 \times 10^{-1} \\ 1.44 \times 10^{-4} \\ 7.60 \times 10^{-6} \\ 8.90 \times 10^{-5} \\ 6.80 \times 10^{-9} \\ 4.94 \times 10^{-6} \end{bmatrix}$$

Step 4 State probability just after Ist proof-test, i.e. at $t = 100$ h and after proof-test.

$$\pi(\tau_{+}) = \Delta e^{\Lambda_{TT}\tau}\pi(0) \quad \tau = T_{\text{proof}}$$

State probability redistribution matrix, Δ as per (4.6) is given as:

$$\Delta = \begin{bmatrix} 1 & 0 & 0 & 0 & 0 & 0 \\ 0 & 1 & 0 & \eta & 0 & 0 \\ 0 & 0 & 1 & 0 & \eta & \eta \\ 0 & 0 & 0 & 1-\eta & 0 & 0 \\ 0 & 0 & 0 & 0 & 1-\eta & 0 \\ 0 & 0 & 0 & 0 & 0 & 1-\eta \end{bmatrix}$$

Probability redistribution matrix, Δ for the example is as:

$$\Delta = \begin{bmatrix} 1 & 0 & 0 & 0 & 0 & 0 \\ 0 & 1 & 0 & 0.9 & 0 & 0 \\ 0 & 0 & 1 & 0 & 0.9 & 0.9 \\ 0 & 0 & 0 & 0.1 & 0 & 0 \\ 0 & 0 & 0 & 0 & 0.1 & 0 \\ 0 & 0 & 0 & 0 & 0 & 0.1 \end{bmatrix}$$

The state probabilities just after Ist proof-test are given as:

$$\pi(100) = \begin{bmatrix} 9.99 \times 10^{-1} \\ 2.25 \times 10^{-4} \\ 1.22 \times 10^{-5} \\ 8.99 \times 10^{-6} \\ 6.86 \times 10^{-10} \\ 4.99 \times 10^{-7} \end{bmatrix}$$

Step 5 State probabilities between Ist and IInd proof-test interval is given by (4.7)

$$\pi(\tau_{+}) = \Delta e^{\Lambda_{TT}\tau}\pi(0) \quad \tau = T_{\text{proof}}$$

and

$$\pi(t) = \pi(\tau + S) = e^{\Lambda_{TT}S}\Delta e^{\Lambda_{TT}\tau}\pi(0) \quad \tau < t < 2\tau; \tau = T_{\text{proof}}$$

Following this step progressively for next proof-test intervals will enable computation of state-probabilities for any specified time.

$$\pi(t) = \pi(n\tau + S) = e^{\Lambda_{TT}S}\alpha^{n}\pi(0) \quad n\tau < t < (n+1)\tau$$

where $\alpha = \Delta e^{\Lambda_{TT}\tau}$.

For operation time $T = 87{,}600$ h, n and S of above equation are 876 and 0, respectively, state probabilities at $t = T$ can be obtained as:

$$\pi(87{,}600) = \begin{bmatrix} 9.20 \times 10^{-1} \\ 6.67 \times 10^{-2} \\ 2.46 \times 10^{-3} \\ 7.39 \times 10^{-3} \\ 5.62 \times 10^{-7} \\ 2.72 \times 10^{-4} \end{bmatrix}$$

Step 6 Once state probabilities with time are known, using mean operator mean probabilities can be obtained:

$$E[\pi(t)] = \bar{\pi} = \frac{\int_0^T \pi(t)\mathrm{d}t}{\int_0^T \mathrm{d}t}$$

$$\bar{\pi} = \frac{1}{T}\left[\Lambda_{TT}^{-1}\left([e^{\Lambda_{TT}\tau} - I][I - \alpha]^{-1}[I - \alpha^n] + [e^{\Lambda_{TT}S} - I]\alpha^n\right)\pi(0)\right]$$

where $T = n\tau + S$.

so for example case it is given as,

$$\bar{\pi} = \begin{bmatrix} 9.99 \times 10^{-1} \\ 1.52 \times 10^{-4} \\ 8.00 \times 10^{-6} \\ 5.49 \times 10^{-5} \\ 4.23 \times 10^{-9} \\ 3.05 \times 10^{-6} \end{bmatrix}$$

Step 7 Determination of mean safety index,

$$mean\,PFaD = 1 - [1_x]\bar{\pi}$$

$$mean\,PFaD = 3.043 \times 10^{-6}$$

Following these steps, state-probabilities and safety index for various proof-test intervals can be estimated.

Case-2 2oo3 architecture. This is similar to previous case. For a given value of proof-test interval, safety index can be evaluated following the described 7 steps.

Step 1 Determination of infinitesimal generator matrix.

From the parameters given in example, Λ_{TT} and Λ_{TA} is given as:

$$\Lambda_{TT} =$$

$$\begin{array}{c} P_{300} \\ P_{210} \\ P_{120} \\ P_{030} \\ P_{201} \\ P_{211} \\ P_{021} \\ P_{102} \\ P_{012} \\ P_{003} \end{array}
\begin{bmatrix}
-2.73\times10^{-5} & 1.25\times10^{-1} & 0 & 0 & 0 & 0 & 0 & 0 & 0 & 0 \\
2.36\times10^{-5} & -1.25\times10^{-1} & 1.25\times10^{-1} & 0 & 0 & 0 & 0 & 0 & 0 & 0 \\
1.99\times10^{-6} & 1.71\times10^{-5} & -1.25\times10^{-1} & 1.25\times10^{-1} & 0 & 0 & 0 & 0 & 0 & 0 \\
2.85\times10^{-7} & 9.50\times10^{-7} & 9.50\times10^{-6} & -1.25\times10^{-1} & 0 & 0 & 0 & 0 & 0 & 0 \\
1.24\times10^{-6} & 0 & 0 & 0 & -1.90\times10^{-5} & 1.25\times10^{-1} & 0 & 0 & 0 & 0 \\
0 & 9.00\times10^{-7} & 0 & 0 & 1.71\times10^{-5} & -1.25\times10^{-1} & 1.25\times10^{-1} & 0 & 0 & 0 \\
0 & 0 & 5.00\times10^{-7} & 0 & 9.50\times10^{-7} & 9.50\times10^{-6} & -1.25\times10^{-1} & 0 & 0 & 0 \\
1.05\times10^{-7} & 0 & 0 & 0 & 9.00\times10^{-7} & 0 & 0 & -3.28\times10^{-5} & 1.25\times10^{-1} & 0 \\
0 & 5.00\times10^{-8} & 0 & 0 & 0 & 5.00\times10^{-7} & 0 & 9.50\times10^{-6} & -1.25\times10^{-1} & 0 \\
1.50\times10^{-8} & 0 & 0 & 0 & 5.00\times10^{-8} & 0 & 0 & 5.00\times10^{-7} & 0 & -2.28\times10^{-5}
\end{bmatrix}$$

Similarly

$$\Lambda_{TA} = \begin{bmatrix} 0 \\ 0 \\ 0 \\ 0 \\ 0 \\ 0 \\ 0 \\ 2.28 \times 10^{-5} \\ 2.28 \times 10^{-5} \\ 2.28 \times 10^{-5} \end{bmatrix}$$

Step 2 Evolution of transient states with time is given by:

$$\dot{\pi}(t) = \Lambda_{TT}\pi(t)$$

$$\pi(t) = \begin{bmatrix} P_{300} \\ P_{210} \\ P_{120} \\ P_{030} \\ P_{201} \\ P_{211} \\ P_{021} \\ P_{102} \\ P_{012} \\ P_{003} \end{bmatrix}$$

Step 3 State probability just before Ist proof-test, i.e. 99 h. (For proof-test interval 100 h)

$$\pi(\tau_-) = e^{\Lambda_{TT}\tau}\pi(0) \quad \tau = T_{\text{proof}}$$

$$\pi(99) = \begin{bmatrix} 9.99 \times 10^{-1} \\ 2.07 \times 10^{-4} \\ 1.82 \times 10^{-5} \\ 2.28 \times 10^{-6} \\ 1.23 \times 10^{-4} \\ 1.78 \times 10^{-8} \\ 9.35 \times 10^{-10} \\ 1.03 \times 10^{-5} \\ 8.08 \times 10^{-10} \\ 1.48 \times 10^{-6} \end{bmatrix}$$

Step 4 State probability just after Ist proof-test, i.e. 100 h. (For proof-test interval 100 h)

$$\pi(\tau_+) = \Delta e^{\Lambda_{TT}\tau}\pi(0) \quad \tau = T_{\text{proof}}$$

State probability redistribution matrix, Δ is given in example, it is again given here.

$$\Delta = \begin{bmatrix} 1 & 0 & 0 & 0 & 0 & 0 & 0 & 0 & 0 & 0 \\ 0 & 1 & 0 & 0 & 0.9 & 0 & 0 & 0 & 0 & 0 \\ 0 & 0 & 1 & 0 & 0 & 0.9 & 0 & 0.9 & 0 & 0 \\ 0 & 0 & 0 & 1 & 0 & 0 & 0.9 & 0 & 0.9 & 0.9 \\ 0 & 0 & 0 & 0 & 0.1 & 0 & 0 & 0 & 0 & 0 \\ 0 & 0 & 0 & 0 & 0 & 0.1 & 0 & 0 & 0 & 0 \\ 0 & 0 & 0 & 0 & 0 & 0 & 0.1 & 0 & 0 & 0 \\ 0 & 0 & 0 & 0 & 0 & 0 & 0 & 0.1 & 0 & 0 \\ 0 & 0 & 0 & 0 & 0 & 0 & 0 & 0 & 0.1 & 0 \\ 0 & 0 & 0 & 0 & 0 & 0 & 0 & 0 & 0 & 0.1 \end{bmatrix}$$

The state probabilities just after Ist proof-test are given as:

$$\pi(100) = \begin{bmatrix} 9.99 \times 10^{-1} \\ 3.18 \times 10^{-4} \\ 2.76 \times 10^{-5} \\ 3.61 \times 10^{-6} \\ 1.23 \times 10^{-5} \\ 1.78 \times 10^{-9} \\ 9.35 \times 10^{-11} \\ 1.03 \times 10^{-6} \\ 8.08 \times 10^{-11} \\ 1.48 \times 10^{-7} \end{bmatrix}$$

Step 5 State probabilities between Ist and IInd proof-test interval is given by (4.7)

$$\pi(\tau_+) = \Delta e^{\Lambda_{TT}\tau}\pi(0) \quad \tau = T_{\text{proof}}$$

and

$$\pi(t) = \pi(\tau + S) = e^{\Lambda_{TT}S}\Delta e^{\Lambda_{TT}\tau}\pi(0) \quad \tau < t < 2\tau; \quad \tau = T_{\text{proof}}$$

Following this step progressively for next proof-test intervals will enable computation of state-probabilities for any specified time.

$$\pi(t) = \pi(n\tau + S) = e^{\Lambda_{TT}S}\alpha^n\pi(0) \quad n\tau < t < (n+1)\tau$$

where $\alpha = \Delta e^{\Lambda_{TT}\tau}$.

For operation time $T = 87{,}600$ h, n and S of above equation are 876 and 0, respectively, state probabilities at $t = T$ can be obtained as:

$$\pi(87{,}600) = \begin{bmatrix} 8.87 \times 10^{-1} \\ 8.88 \times 10^{-2} \\ 5.30 \times 10^{-3} \\ 6.60 \times 10^{-4} \\ 9.85 \times 10^{-3} \\ 1.42 \times 10^{-6} \\ 7.49 \times 10^{-8} \\ 5.85 \times 10^{-4} \\ 4.45 \times 10^{-8} \\ 7.30 \times 10^{-5} \end{bmatrix}$$

Step 6 Once state probabilities with time are known, using mean operator mean probabilities can be obtained:

$$E[\pi(t)] = \bar{\pi} = \frac{\int_0^T \pi(t)\mathrm{d}t}{\int_0^T \mathrm{d}t}$$

$$\bar{\pi} = \frac{1}{T}\left[\Lambda_{TT}^{-1}\left([e^{\Lambda_{TT}\tau} - I][I - \alpha]^{-1}[I - \alpha^n] + [e^{\Lambda_{TT}S} - I]\alpha^n\right)\pi(0)\right]$$

where $T = n\tau + S$.

So for example case it is given as,

$$\bar{\pi} = \begin{bmatrix} 9.99 \times 10^{-1} \\ 2.18 \times 10^{-4} \\ 1.92 \times 10^{-5} \\ 2.40 \times 10^{-6} \\ 7.60 \times 10^{-5} \\ 1.11 \times 10^{-8} \\ 5.80 \times 10^{-10} \\ 6.41 \times 10^{-6} \\ 5.09 \times 10^{-10} \\ 9.15 \times 10^{-7} \end{bmatrix}$$

Step 7 Determination of mean safety index,

$$mean\ PFaD = 1 - [1_x]\bar{\pi}$$

$$mean\ PFaD = 7.321 \times 10^{-6}$$

Following these steps, state-probabilities and safety index for various proof-test interval can be derived.

4.6.2.3 Discussion

For the systems with identical channels *PFaD* values of *1oo2* architecture is lower than *2oo3* architecture for all proof-test intervals. *PFaD* values for both the architecture increases with increase in proof-test interval. Manifested availability values of *2oo3* architecture are high compared with *1oo2*. From Fig. 4.16 it can be observed that there is no appreciable decrease in availability of *2oo3* architecture with proof-test interval. While *1oo2* architecture shows decrease in availability with increasing proof-test interval.

One factor for higher *PFaD* value of *1oo2* architecture is it spends more time (compared with *2oo3*) in F_S (i.e. safe shutdown). Lower value of availability for *1oo2* proves this fact.

4.7 Advantage of Modeling Safe Failures

Safety index $PFaD(t)$ along with availability index *mAv(t)* can be very useful at the time of safety system design as well as operation phase. During design phase, these two indices can be evaluated for different design alternatives (architecture, hazard rates, DC, CCF) with specified external factors (Proof-test interval, MTBD) for a specified time. The design alternative which gives lowest $PFaD(t)$ (lower than required) and maximum *mAv(t)* is best design option.

Table 4.7 System parameter values for comparison

Parameter	Case-I	Case-II	Case-III
λ_S	$2 \times 10^{-6}\ h^{-1}$	$4.45 \times 10^{-5}\ h^{-1}$	$4.45 \times 10^{-6}\ h^{-1}$
λ_D	$2 \times 10^{-6}\ h^{-1}$	$4.10 \times 10^{-5}\ h^{-1}$	$4.10 \times 10^{-6}\ h^{-1}$
DC	0.9	0.99	0.99
η	1	1	1
β	0	0	0
β_2	0	0	0
μ	$1/8\ h^{-1}$	$1\ h^{-1}$	$1\ h^{-1}$
T		10,000 h	
MTBD		[5,000, 7,500] h	
T_{Proof}		2,000 h	

Table 4.8 *PFaD* and *mAv* values for comparison

MTBD	Case-I		Case-II		Case-III	
	5,000 h	7,500 h	5,000 h	7,500 h	5,000 h	7,500 h
(a) PFaD values						
1oo2	4.36E-08	3.11E-08	1.82E-07	1.25E-07	3.19E-08	1.09E-07
2oo3	1.31E-07	8.97E-08	5.50E-07	3.78E-07	3.33E-08	5.29E-08
(b) Availability values						
1oo2	0.99993666	0.99993665	0.99982906	0.99982906	0.99998292	0.99998292
2oo3	0.99999999	0.99999999	0.99999996	0.99999996	1	1

An example with three cases is taken, system parameter values along with common environment parameter values foe all three are given in Table 4.7. Both *PFaD* and availability are evaluated for all the 3 cases for *1oo2* and *2oo3* architectures at two different values of MTBD. The results are shown in Table 4.8.

Case-I with *1oo2* architecture and case-III with *2oo3* architecture gives lowest value of *PFaD* for both MTBD values compared to others. Now looking at availability values gives that availability is highest. So, case-III with *2oo3* architecture seems to be preferable.

During operational phase of system mainly proof-test interval is tuned to achieve the target safety. Frequent proof-tests increase safety but decrease the availability of safety system as well as of EUC. These two indices are helpful in deciding the maximum proof-test interval which will met required $PFaD(t)$ value with maximizes $mAv(t)$.

4.8 DSPN Based Safety Models

The safety model discussed here, require Markov models to be manually made and various matrices to be deduced from the system parameters. Stochastic Petri based tools- SPNP [48], TimeNET [49] proves to be helpful as they provide a graphical interface to specify the problem and numerically gives desired measure. The safety model discussed here, have a deterministic event- periodic proof-test. A class of SPN called DSPN can model and solve systems with combination of exponential and deterministic events. DSPNs have some limitations, a detail overview of Petri net based tools is given in [48, 49].

DSPN based safety models of *1oo2* and *2oo3* system architectures are shown in Figs. 4.17 and 4.18, respectively.

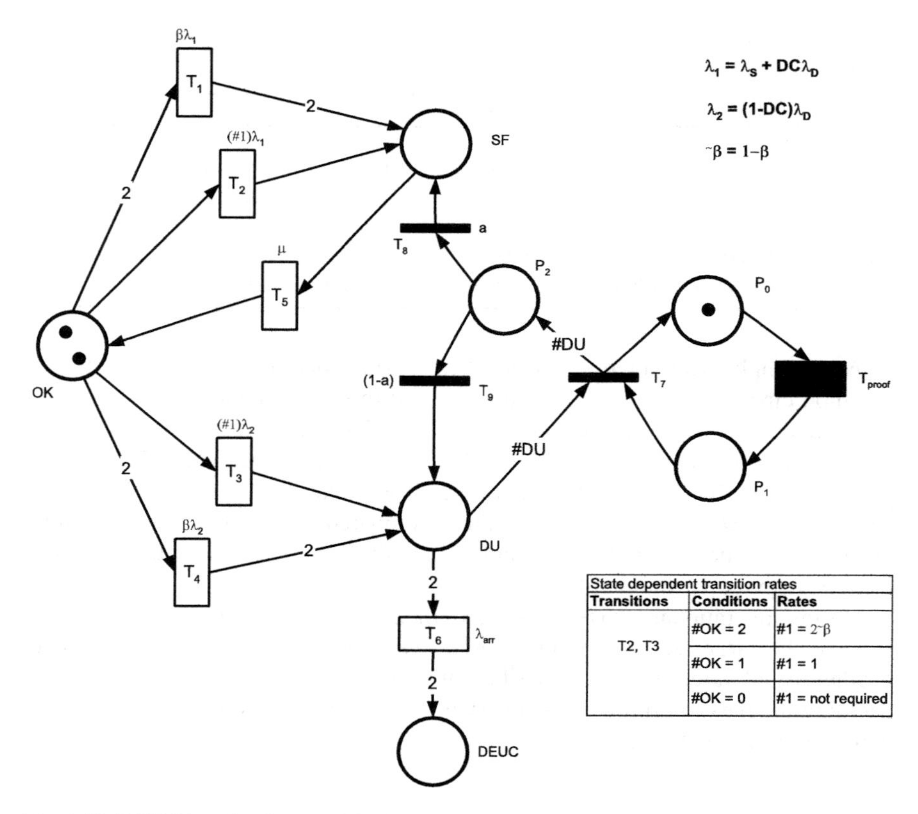

State dependent transition rates		
Transitions	Conditions	Rates
T2, T3	#OK = 2	#1 = 2¯β
	#OK = 1	#1 = 1
	#OK = 0	#1 = not required

Fig. 4.17 DSPN based safety model of *1oo2* system. Number of tokens in places *OK*, *SF* and *DU* represent number of channels in healthy state, safe failure and dangerous failure state respectively. Transition T_1 (T_4) represents safe (dangerous) hazard rate due to CCF. Transition T_2 (T_3) represents safe (dangerous) hazard rate without CCF. T_5 depicts the repair rate of channel from safe failure state. Demand arrival rate is shown with transition T_6. Places P_0 and P_1 along with deterministic transition T_{proof} and immediate transition T_7 model periodic proof-test. When token is in place P_0, transition T_{proof} is enabled and fires after a deterministic time. On firing a token is deposited in place P_1. In this marking immediate transition T_7 become enable and fires immediately and removes all the tokens from place *DU*. All the token of *DU* are deposited in place P_2 and one token is deposited in P_0. From P_2 tokens may go to place *SF* or *DU* based on degree of proof-test

4.9 Summary

For computer-based systems, applicable dependability attributes depends on its usage application area. If a computer-based system is used in safety-critical application, dependability attribute *safety* is the most appropriate. Similarly, for mission-critical and economically-critical application reliability and availability, respectively, are the most appropriate. A brief survey of reliability, availability and

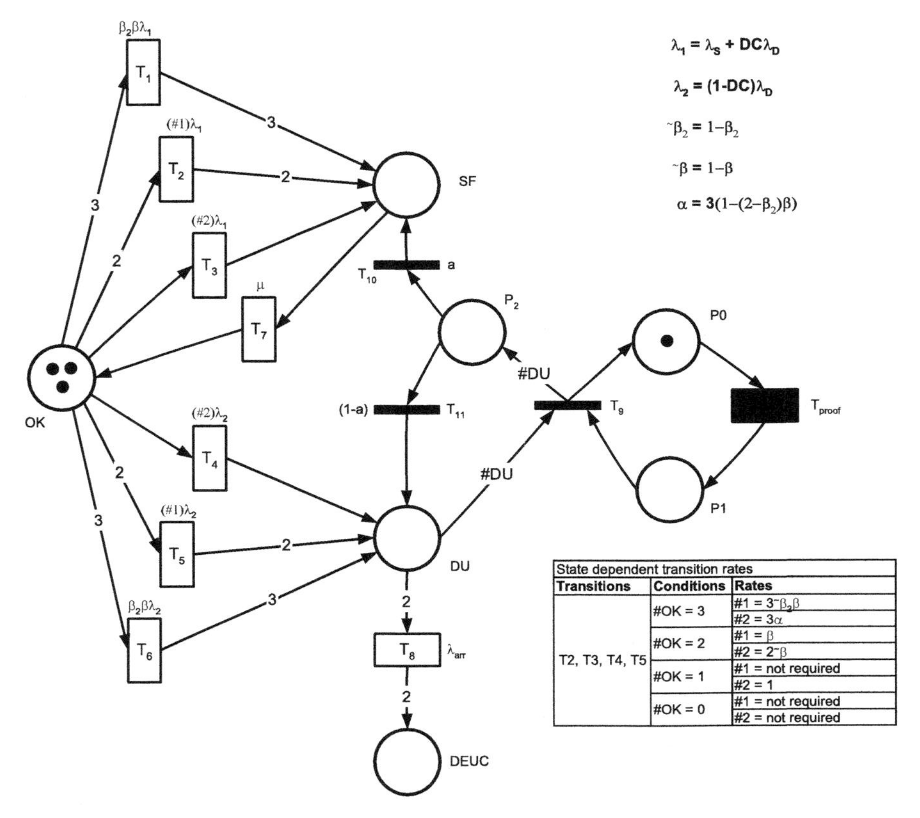

Fig. 4.18 DSPN based safety model of *2oo3* system. Number of tokens in places *OK*, *SF* and *DU* represent number of channels in healthy state, safe failure and dangerous failure state respectively. Transition T_1, T_2 (T_5, T_6) represents safe (dangerous) hazard rate due to CCF. Transition T_3 (T_4) represents safe (dangerous) hazard rate without CCF. T_7 depicts the repair rate of channel from safe failure state. Demand arrival rate is shown with transition T_8. Places P_0 and P_1 along with deterministic transition T_{proof} and immediate transition T_9 model periodic proof-test. When token is in place P_0, transition T_{proof} is enabled and fires after a deterministic time. On firing a token is deposited in place P_1. In this marking immediate transition T_9 become enable and fires immediately and removes all the tokens from place *DU*. All the token of *DU* are deposited in place P_2 and one token is deposited in P_0. From P_2 tokens may go to place *SF* or *DU* based on degree of proof-test

safety models from literature is given. Safety models of IEC 61508 has been extended to incorporate demand rate, the method is illustrated in detail.

Quantitative safety index *PFD* is published in safety standard IEC 61508. Various researchers have contributed to make the method more clear, usable and relevant. Contributing in the same direction Markov model for the systems considering safe failures, periodic proof-tests (perfect as well as imperfect) and demand rate have been derived. The analysis has been done to derive closed form solution for performance based safety index *PFaD* and availability. The advantages of modeling safe failures are shown with the help of an example.

Reliable data on process demands is needed to correctly estimate demand rate and its distribution.

References

1. Avizienis A, Laprie J-C, Randell B (2000) Fundamental concepts of dependability. In: Proceeding of 3rd Information Survivability Workshop, pp 7–11, October 2000
2. Johnson BW (1989) Design and analysis of fault-tolerant digital systems. Addison Wesley, New York
3. Lala PK (1985) Fault tolerant and fault testable hardware design. PHI
4. Zang X, Sun H, Trivedi KS (1999) A BDD-based algorithm for reliability graph analysis. IEEE Trans Reliab 48(1):50–60
5. Zang X (1999) Dependability modeling of computer systems and networks. PhD thesis, Department of Electrical and Computer Engineering, Duke University
6. Dugan JB, Bavso SJ, Boyd MA (1992) Dynamic fault-tree models for fault-tolerant computer systems. IEEE Trans Reliab 41(3):362–377
7. Trivedi KS (1982) Probability & statistics with reliability, queueing, and computer science applications. Prentice-Hall, Englewood Cliffs
8. Mishra KB (1992) Reliability analysis and prediction. Elsevier, Amsterdam
9. Goseva-Popstojanova K, Trivedi KS (2001) Architecture-based approach to reliability assessment of software systems. Performance Evaluation 45(2–3):179–204
10. Pham H (2000) Software reliability. Springer, Berlin
11. Pham H (1996) A software cost model with imperfect debugging random life cycle and penalty cost. Int J Syst Sci 25(5):455–463
12. Goel AL (1985) Software reliability models: Assumptions, limitations, and applicability. IEEE Trans Softw Eng SE-2(12):1411–1423
13. Littlewood B (1975) A reliability model for systems with markov structure. Appl Stat 24(2):172–177
14. Cheung RC (1980) A user-oriented software reliability model. IEEE Trans Softw Eng 6(2):118–125
15. Laprie JC (1984) Dependability evaluation of software systems in operation. IEEE Trans Softw Eng 10(6):701–714
16. Kubat P (1989) Assessing reliability of modular software. Oper Res Lett 8:35–41
17. Gokhale SS, Trivedi KS (2006) Analytical models for architecture-based software reliability prediction: a unification framework. IEEE Trans Reliab 55(4):578–590
18. Gokhale SS, Trivedi KS (1999) A time/structure based software reliability model. Ann Softw Eng 8:85–121
19. Ledoux J (1999) Availability modeling of modular software. IEEE Trans Softw Eng 48(2):159–168
20. Shooman M (1976) Structural models for software reliability prediction. In: Proceeding of 2nd International Conference on Software Engineering. San Francisco, CA, pp 268–280
21. Yacoub S, Cukic B, Ammar HH (2004) A scenario-based reliability analysis approach for component-based software. IEEE Trans Reliab 53(4):465–480
22. Xie M, Wohlin C (1995) An additive reliability model for the analysis of modular software failure data. In: Proceedings of the 6th International Symposium on Software Reliability Engineering (ISSRE'95), Toulouse, France, pp 188–194

23. Everett W (1999) Software component reliability analysis. In: Proceeding of the symposium on Application-Specific Systems and Software Engineering Technology (ASSET'99), Dallas, TX, pp 204–211
24. IEC 61508: Functional safety of electric/electronic/programmable electronic safety-related systems, Parts 0–7; October 1998–May 2000
25. Bukowski JV (2001) Modeling and analyzing the effects of periodic inspection on the performance of safety-critical systems. IEEE Trans Reliab 50(3):321–329
26. Guo H, Yang X (2007) A simple reliability block diagram method for safety integrity verification. Reliab Eng Syst Saf 92:1267–1273
27. Zhang T, Long W, Sato Y (2003) Availability of systems with self-diagnostics components-applying markov model to IEC 61508-6. Reliab Eng Syst Saf 80:133–141
28. Bukowski JV, Goble WM (2001) Defining mean time-to-failure in a particular failure-state for multi-failure-state systems. IEEE Trans Reliab 50(2):221–228
29. Brown S (2000) Overview of IEC 61508: functional safety of electrical/electronic/ programmable electronic safety-related systems. Comput Control Eng J 11(1):6–12
30. Bukowski JV (2005) A comparison of techniques for computing PFD average. In: RAMS 2005, pp 590–595
31. Goble WM, Bukowski JV (2001) Extending IEC61508 reliability evaluation techniques to include common circuit designs used in industrial safety systems. In: Proceeding of Annual Reliability and Maintainability Symposium, pp 339–343
32. Hokstad P, Carneliussen K (2004) Loss of safety assesment and the IEC 61508 standard. Reliab Eng Syst Saf 83:111–120
33. Summers A (2000) Viewpoint on ISA TR84.0.02-simplified methods and fault tree analysis. ISA Trans 39(2):125–131
34. Scherrer C, Steininger A (2003) Dealing with dormant faults in an embedded fault-tolerant computer system. IEEE Trans Reliab 52(4):512–522
35. Delong TA, Smith T, Johnson BW (2005) Dependability metrics to assess safety-critical systems. IEEE Trans Reliab 54(2):498–505
36. Choi CY, Johnson RW, Profeta JA III (1997) Safety issues in the comparative analysis of dependable architectures. IEEE Trans Reliab 46(3):316–322
37. Bukowski JV (2006) Incorporating process demand into models for assessment of safety system performance. In: RAMS 2006, pp 577–581
38. Rouvroye JL, Brombacher AC (1999) New quantitative safety standards: different techniques, different results? Reliab Eng Syst Saf 66:121–125
39. Manoj K, Verma AK, Srividya A (2007) Analyzing effect of demand rate on safety of systems with periodic proof-tests. Int J Autom Comput 4(4):335–341
40. Manoj K, Verma AK, Srividya A (2008) Modeling of demand rate and imperfect proof-test and analysis of their effect on system safety. Reliab Eng Syst Saf 93:1720–1729
41. Manoj K, Verma AK, Srividya A (2008) Incorporating process demand in safety evaluation of safety-related systems. In: Proceeding of Int Conf on Reliability, Safety and Quality in Engineering (ICRSQE-2008), pp 378–383
42. Cox DR, Miller HD (1970) The theory of stochastic processes. Methuen & Co, London
43. Darroch JN, Seneta E (1967) On quasi-stationary distributions in absorbing continuous-time finite markov chains. J Appl Probab 4:192–196
44. Marsan MA, Chiola G (1987) On petri nets with deterministic and exponentially distributed firing times. In: Advances in Petri Nets 1986, Lecture Notes in Computer Science 266, pp 132–145
45. Varsha M, Trivedi KS (1994) Transient analysis of real-time systems using deterministic and stochastic petri nets. In: Int'l Workshop on Quality of Communication-Based Systems
46. Khobare SK, Shrikhande SV, Chandra U, Govidarajan S (1998) Reliability analysis of micro computer modules and computer based control systems important to safety of nuclear power plants. Reliab Eng Syst Saf 59(2):253–258

47. Khobare SK, Shrikhande SV, Chandra U, Govidarajan G (1995) Reliability assessment of standardized microcomputer circuit boards used in C&I systems of nuclear reactors. Technical report BARC/1995/013
48. Trivedi KS (2001) SPNP user's manual, version 6.0. Technical report
49. Zimmermann A, and Knoke M (2007) TimeNET 4.0 user manual. Technical report, August 2007

Chapter 5
Network Technologies for Real-Time Systems

5.1 Introduction

The purpose of this chapter is to introduce basic term and concepts of network technology. Main emphasis of is on schedulers and real-time analysis of these networks. Networks used in critical applications, such as, CAN and MIL-STD-1553B are discussed in detail.

5.2 Network Basics

Network aims at providing reliable, timely and deterministic communication of data between connected devices. The communication is carried out over a communication network relying on either a wired or a wireless medium.

To manage the complexity of communication protocol, reference model have been proposed such as ISO/ISO layers [1]. The model contains seven layers—Application layer, Presentation layer, Session layer, Transport layer, Network layer, Data link layer, and Physical layer. The lowest three layers—Network, Data link and Physical—are networks dependent, the physical layer is responsible for the transmission of raw data on the medium used. The data link layer is responsible for the transmission of data frames and to recognise and correct errors related to this. The network layer is responsible for the setup and maintenance of network wide connections. The upper three layers are application oriented, and the intermediate layers (transport layer) isolates the upper three and the lower three layers from each other, i.e. all layers above the transport layer can transmit messages independent of the underlying network infrastructure.

In this book, the lower layers of the ISO/OSI reference model are of great importance, where for real-time communications, the Medium Access Control (MAC) protocol determines the degree of predictability of the network technology.

A. K. Verma et al., *Dependability of Networked Computer-based Systems*,
Springer Series in Reliability Engineering, DOI: 10.1007/978-0-85729-318-3_5,

Usually, the MAC protocol is considered a sub layer of the physical layer or the data link layer.

5.3 Medium Access Control (MAC) Protocols

In a node with networking capabilities a local communication adapter mediates access to the medium used for message transmission. Upper layer application that sends messages send them to the local communication adapter. Then, the communication adapter takes care of the actual message transmission. Also, the communication adapter receives messages from the medium. When data is to be sent from the communication adapter to the physical medium, the message transmission is controlled by the medium access control protocols (MAC protocols).

MAC protocol are mainly responsible for variation in end-to-end delay times. Widely used MAC protocols can be classified as follows [2]:

- random access protocols, examples are,
  - CSMA/CD (Carrier Sense Multiple Access/Collision Detection)
  - CSMA/CR (Carrier Sense Multiple Access/Collision Resolution)
  - CSMA/CA (Carrier Sense Multiple Access/Collision Avoidance)
- fixed-assignment protocols, examples are,
  - TDMA (Time Division Multiple Access)
  - FTDMA (Flexible TDMA)
- demand-assignment protocols, examples are,
  - distributed solutions relying on tokens
  - centralised solutions by the usage of masters

These MAC protocols are used for both real-time and non real-time communications, and each of them have different timing characteristics.

5.3.1 Carrier Sense Multiple Access/Collision Detection (CSMA/CD)

In carrier sense multiple access/collision detection (CSMA/CD) collision may occur but it relies on detection of collision. In the network collisions between messages on the medium are detected by simultaneously writing the messages on the medium are detected by simultaneously writing the message and reading the transmitted signal on the medium. Thus, it is possible to verify if the transmitted signal is the same as the signal currently being transmitted. If both are not the same, one or more parallel transmissions are going on. Once a collision is detected the transmitting stations stop their transmissions and wait for some time (i.e. backoff) before retransmitting the message in order to reduce the risk of the same messages colliding again. However, due to the possibility of successive collisions, the temporal behavior of CSMA/CD networks can be somewhat hard to predict. 1-persistent CSMA/CD is used for Ethernet [3].

5.3.2 Carrier Sense Multiple Access/Collision Resolution (CSMA/CR)

Operation of carrier sense multiple access/collision resolution (CSMA/CR) is similar to CSMA/CD. The difference is CSMA/CR does not go into backoff mode once there is a collision detected. Instead, CSMA/CR resolves collisions by determining one of the message transmitters involved in the collision that is allowed to go on with an uninterrupted transmission of its messages. The other message(s) involved in the collision are retransmitted at another time. Due to the collision resolution feature of CSMA/CR, it has the possibility to become more predictable in its temporal behavior compared to CSMA/CD. An example of a network technology that implements CSMA/CR is CAN [22].

5.3.3 Carrier Sense Multiple Access/Collision Avoidance (CSMA/CA)

In some cases it is not possible to detect collisions although it might still be desirable to try to avoid them. For example, using a wireless medium often makes it impossible to simultaneously read and write (send and receive) to the medium, as (at the communication adapter) the signal sent is so much stronger than (and therefore overwrites) the signal received. Carrier Sense Multiple Access/Collision Avoidance (CSMA/CA) protocols can avoid collisions by the usage of some handshake protocol in order to guarantee a free medium before the initiation of a message transmission. CSMA/CA is used by ZigBee [3].

5.3.4 Time Division Multiple Access (TDMA)

Time Division Multiple Access (TDMA) is a fixed assignment MAC protocol where time is used to achieve temporal partitioning of the medium. Messages are sent at predetermined instances in time, called message slots. Often, a schedule of slots or exchange table is prepared, and this schedule is then followed and repeated during runtime.

Due to the time slotted nature of TDMA networks, their temporal behavior is very predictable and deterministic. TDMA networks are therefore very suitable for safety-critical systems with hard real-time guarantees. A drawback of TDMA networks is that they are not flexible, as messages can not be sent at an arbitrary time and changing message table is somewhat difficult. A message can only be sent in one of the message's predefined slots, which affect the responsiveness of the message transmissions. Also, if a message is shorter than its allocated slot, bandwidth is wasted since the unused portion of the slot cannot be used by another message. Example of TDMA real-time network is MIL-STD-1553B [4] and TTP/C [5]. In both these, exchange tables are created offline. One example of an online scheduled TDMA network is the GSM network.

5.3.5 Flexible Time Division Multiple Access (FTDMA)

Another fixed assignment MAC protocol is Flexible Time Division Multiple Access (FTDMA). As regular TDMA networks, FTDMA networks avoid collisions by dividing time into slots. However, FTDMA networks use mini slotting concept in order to make more efficient use of bandwidth, compared to TDMA network. FTDMA is similar to TDMA with the difference in run-time slot size. In an FTDMA schedule the size of a slot is not fixed, but will vary depending on whether the slot is used or not. In case all slots are used in a FTDMA schedule, FTDMA operated the same way as TDMA. However, if a slot is not used within a small time offset after its initiation, the schedule will progress to its next slot. Hence, unused slots will be shorter compared to a TDMA network where all slots have fixed size. However, used slots have the same size in both FTDMA and TDMA networks. Variant of FTDMA can be found in Byteflight [6], and FlexRay [7].

5.3.6 Distributed Solutions Relying on Tokens

An alternative way of eliminating collisions on the network is to achieve mutual exclusion by the usage of token based demand assignment MAC protocols. Token based MAC protocols provide a fully distributed solution allowing for exclusive usage of the communications networks to one transmitter at a time.

In token (unique within the network) networks only the owner of the token is allowed to transmit messages on the network. Once the token holder is done with transmitting messages, or has used its alloted time, the token is passed to another node. Examples of the protocols are TTP (Timed Token Protocol) [8], IEEE 802.5 Token Ring Protocol, IEEE 802.4 Token Bus Protocol and PROFIBUS [9, 10].

5.3.7 Master/Slave

Another example of demand assignment MAC protocols is the centralised solution relying on a specialised node called the master node. The other nodes in the system are called slave nodes. In master/slave networks, elimination of message collisions is achieved by letting the master node control the traffic on the network, deciding which messages are allowed to be sent and when. This approach is used in TTP/A [5, 11] and PROFIBUS [9, 10].

5.4 Networks

Communication network technologies are either wired networks or wireless. The medium can be either wired, transmitting electrical or optical signals in cables or optical fibres, or wireless, transmitting radio signals or optical signals. In this text, we will constraint ourself to wired networks only.

5.4.1 Ethernet

In parallel with the development of various fieldbus technologies providing real-time communication for avionics, trains, industrial and process automation, and building and home automation, Ethernet established itself as the de facto standard for non real-time communications. Comparing networking solutions for automation networks and office networks, fieldbuses were initially the choice for DCCSs and automation networks. At the same time, Ethernet evolved as the standard for office automation, due to its popularity, prices on Ethernet based networking solutions dropped. A lower price on Ethernet controllers made it interesting to develop additions and modifications to Ethernet for real-time communications, allowing Ethernet to compete with established real-time networks.

Ethernet is not very suitable to real-time communication due to its handling of message collisions. DCCSs and automation networks require timing guarantees for individual messages. Several proposals to minimise or eliminate the occurrence of collisions on Ethernet have been proposed over the years. The stronger candidate today is the usage of a switch based infrastructure, where the switches separate collision domains to create a collision free network providing real-time message transmissions over Ethernet [12, 13].

Other proposals providing real-time predictability using Ethernet include, making Ethernet more predictable using TDMA [14], offline scheduling or token algorithms [15]. Note that a dedicated network is usually required when using tokens, where all nodes sharing the network must obey the token protocol [8]. A different approach for predictability is to modify the collision resolution algorithm.

Other predictable approaches are usage of a master/slave concept as FTT-Ethernet [16], or the usage of Virtual Time CSMA (VTCSMA) [17] protocol, where packets are delayed in a predictable way in order to eliminate the occurrence of collisions. Moreover, window protocols [18] are using a global window (synchronized time interval) that also remove collisions. The window protocol is more dynamic and somewhat more efficient in its behavior compared to the VTCSMA approach.

Without modification to the hardware or networking topology (infrastructure), the usage of traffic smoothing [19, 20] can eliminate bursts of traffic, which have severe impact on the timely delivery of message packets on the Ethernet. By keeping the network load below a given threshold, a probabilistic guarantee of message delivery can be provided. Some more detail about the Ethernet is given below:

Ethernet is the most widely used local area networking (LAN) technology for home and office use in the world today. Ethernet is in existence for almost 3 decades. A brief history of evolution of Ethernet over the years is given below.

5.4.1.1 Evolution of Ethernet

Ethernet network system was invented for interconnecting advanced computer workstations, making it possible to send data to one another and to high-speed laser printers. It was invented at the Xerox Palo Alto Research Center, USA by Bob Metcalfe in 1973. To make the wide spread use of Ethernet in market and interoperability, a need to standardize Ethernet was felt. Due to initiative of DEC-Intel-Xerox (DIX), the first 10Mbps Ethernet was published in 1980. The standard, entitled *The Ethernet, A Local Area Network: Data Link Layer and Physical Layer Specifications*, contained specifications for operation as well as media. After the DIX standard was published, a new effort by IEEE to develop open network standard also started. The IEEE standard was formulated under the direction of IEEE Local and Metropolitan Networks (LAN/MAN) Standards Committee, which identifies all the standards it develop with number 802. There has been a number of networking standards published in the 802 branch of IEEE, including the 802.3* Ethernet and 802.5 Token Ring Standards.

5.4.1.2 CSMA/CD in Ethernet

Carrier Sense Multiple Access with Collision Detection is base-band multiple access technique. It outlines a complete algorithm [1]. Ethernet is based on CSMA/CD, it is 1-persistence CSMA/CD with exponential backoff algorithm. Simplified transmission protocol can be specified as follows:

- No slots
- Node with data ready for transmission checks the medium whether it is idle or busy. This is called *carrier sense*.
- Transmitting node aborts when it senses that another node is transmitting, that is, *collision detection.*
- before attempting a retransmission, adapter waits a random time.

The transmission procedure can be described as:

```
get a datagram from upper layer
  K := 0; n:=0;
  repeat:
  wait for K*512 bit-time;
  while (bus busy) wait;
  wait for 96 bit-time after detecting no signal;
  transmit and detect collision;
  if detect collision
  stop and transmit a 48-bit jam;
  n++;
  m:=min(n,10), where n is the number of collisions
  choose K randomly from 0, 1, 2, ..., 2m - 1
  if n < 16 goto repeat
  else giveup
  else done!
```

5.4.1.3 Switched Ethernet

Switched Ethernet is a full duplex operation of Ethernet. This capability allows simultaneous two-way transmission over point-to-point links. This transmission is functionally much simpler than half-duplex transmission because it involves no media contention, no collisions, no need to schedule retransmissions, and no need for extension bits on the end of short frames. The result is not only more time available for transmission, but also an effective doubling of the link bandwidth because each link can now support full-rate, simultaneous, two-way transmission.

Transmission can usually begin as soon as frames are ready to send with only restriction of minimum inter frame gap between successive frames as per Ethernet frame standards.

When sending a frame in full-duplex mode, the station ignores carrier sense and does not defer to traffic being received on the channel. But it waits for the inter frame gap. Providing inter frame gap ensures that the interfaces a each end of the link can keep up with the full frame rate of the link. In full-duplex mode, the stations at each end of the link ignore any collision detect signals that come from the transceiver. The CSMA/CD algorithm used on shared half-duplex Ethernet channels is not used on a link operating in full-duplex mode. A station on a full-duplex link sends whenever it likes, ignoring carrier sense (CS). There is no multiple access (MA) since there is only one station at each end of the link and the Ethernet channel between them is not the subject of access contention by multiple stations. Since there is no access contention, there will be no collision either, so the station at each end of the link is free to ignore collision detect (CD).

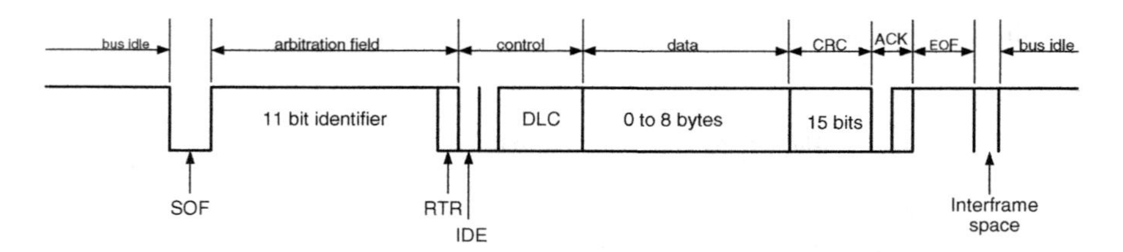

Fig. 5.1 CAN message format. It does not have source/destination address. 11-bit identifier is used for filtering at receiver as well as to arbitrate access to bus. It has 7 control bits- RTR (remote retransmission request), IDE (identifier extension), reserve bit for future extensions and 4 bits to give length of data field, DLC (data length code). Data field can be 0 to 8 bytes in length and CRC field contain 15-bit code that can be used to check frame integrity. Following CRC field is acknowledge (ACK) field comprising an ACK slot bit and an ACK delimiter bit

5.4.2 Controller Area Network (CAN)

Controller Area Network (CAN) is a broadcast bus- a single pair of wires- where a number of nodes are connected to the bus. It employs carrier sense multiple access with collision detection and arbitration based on message priority (CSMA/AMP) [21]. The basic features [21, 22] of CAN are:

1. High-speed serial interface: CAN is configurable to operate from a few kilobits to 1 Mega bits per second.
2. Low cost physical medium: CAN operates over a simple inexpensive twisted wire pair.
3. Short data lengths: The short data length of CAN messages mean that CAN has very low latency when compared to other systems.
4. Fast reaction times: The ability to transmit information without requiring a token or permission from a bus arbitrator results in extremely fast reaction times.
5. Multi master and peer-to-peer communication: Using CAN it is simple to broadcast information to all or a subset nodes on the bus and just an easy to implement peer-to-peer communication.
6. Error detection and correction: The high level of error detection and number of error detection mechanisms provided by the CAN hardware means that CAN is extremely reliable as a networking solution.

Data is transmitted as *message*, consisting of up to 8 bytes. Format of CAN message set is shown in Fig. 5.1. Every message is assigned a unique *identifier*. The identifier serves two purposes, filtering messages upon reception and assigning priority to the message.

The use of identifier as priority is the most important part of CAN regrading real-time performance. The identifier field of CAN message is used to control access to the bus after collision by taking advantage of certain electrical characteristics. In case of multiple stations transmitting simultaneously, all stations will see 0 if any one of the node puts 0 bit (dominant), while all stations will see 1 if all transmitting node put 1 bit. So, during arbitration, by monitoring the bus a node

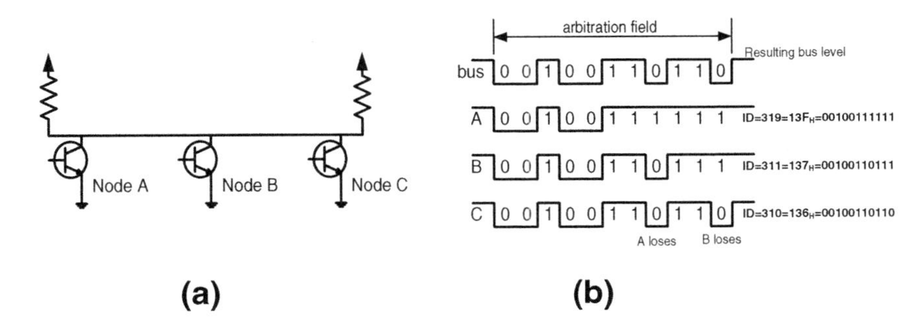

Fig. 5.2 **a** CAN's electrical interface (wired-OR) which enables priority based arbitration, **b** arbitration mechanism when 3 nodes are transmitting simultaneously

detects if there is a competing higher priority message and stops transmission if this is the case. A node transmitting the last bit of the identifier without detecting a higher priority message must be transmitting the highest priority ready message, and hence can continue. Figure 5.2 shows the arbitration mechanism and electrical diagram of CAN.

CAN uses principle of hard synchronization [22]. So, to allow receivers to synchronize and adjust internal timing, CAN insert a bit of opposite polarity when 5 consecutive bit of same polarity are transmitted on the bus. This process is called *bit-stuffing* and bits inserted by this process are called *stuff-bits*. The stuff-bits are removed at the receiver. Bit-stuffing affects the transmission time of message.

For more details on CAN, interested readers are requested to refer [21, 22].

5.4.3 MIL-STD-1553B

MIL-STD-1553B is a military standard that defines the electrical and protocol characteristics for a data bus. The data bus is used to provide a medium for exchange of data and information between various nodes of a system. This standard defines requirement for digital, command/response, time division multiplexing techniques for a 1 MHz serial data bus and specifies the data bus and its interface electronics [4]. Originally this standard was intended for Air Force applications. But with its wide acceptance and usage, it is being used in a large number of critical applications, such as, space shuttles, space stations, surface ships, submarines, helicopters, tanks, subways and manufacturing production lines.

A summary of the characteristics of MIL-STD-1553B is given in Table 5.1 [23]. The standard defines four hardware elements. These are:

1. Transmission media
2. Remote terminals
3. Bus controllers
4. Bus monitors

Table 5.1 Characteristics of MIL-STD-1553B

Data rate	1 MHz
Word length	20 bits
Data bits/word	16 bits
Message length	Maximum of 32 data words
Transmission technique	Half-duplex
Operation	Asynchronous
Encoding	Manchester II bi-phase
Protocol	Command/response
Bus control	Single or multiple
Fault tolerance	Typically dual redundant, second bus in "Hot Backup" status
Message formats	1. Bus controller to terminal
	2. Terminal to bus controller
	3. Terminal to terminal
	4. Broadcast
	5. System control
Number of remote terminals	Maximum of 31
Terminal types	1. Remote terminals
	2. Bus controller
	3. Bus monitor
Transmission media	Twisted shielded pair
Coupling	Transformer and direct

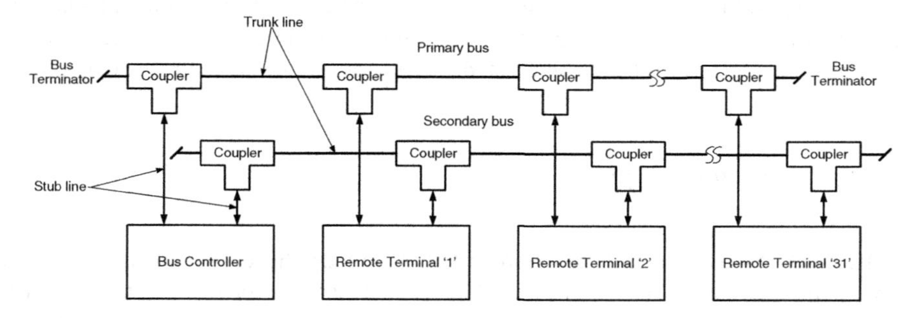

Fig. 5.3 MIL-STD-1553B network

A typical network consisting of a bus controller and a remote terminal with dual redundant bus is shown in Fig. 5.3.

The control, data flow, status reporting, and management of the bus are provided by three word types:

1. Command words
2. Data words
3. Status words

Word formats are shown in Fig. 5.4.

The primary purpose of the data bus is to provide a common media for the exchange of data between terminals of system. The exchange of data is based on

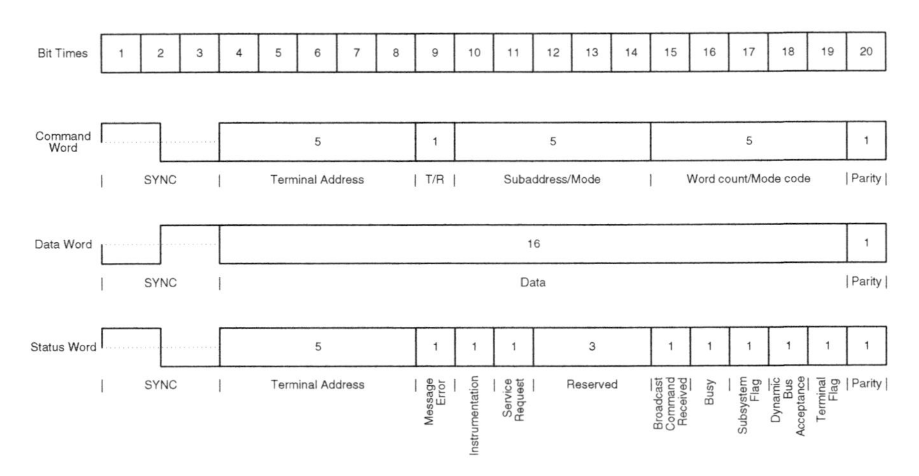

Fig. 5.4 Messages formats

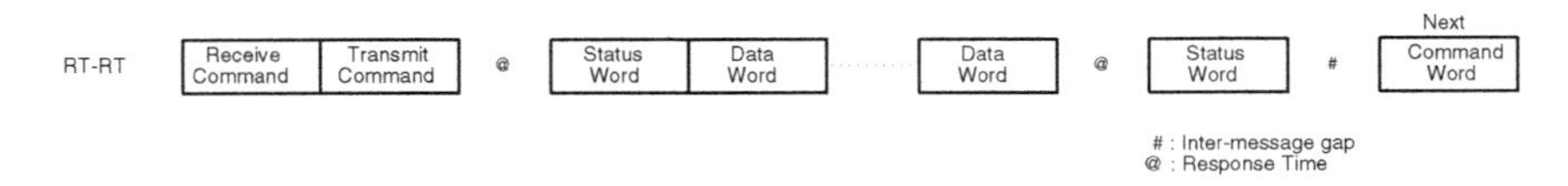

Fig. 5.5 RT-RT information transfer format

message transmission formats. The standard defines ten types of message transmission formats. All of these formats are based on the three word types defined in succeeding paragraph. A RT-RT information transfer format is shown in Fig. 5.5.

Intermessage gap shown in Fig. 5.5, is the minimum gap time that the bus controller shall provide between messages. Its typical value is 4 μs. Response time is time period available for terminals to respond to a valid command word. This period is of 4–12 μs. A time out occurs if a terminal do not respond within No-Response timeout, it is defined as the minimum time that a terminal shall wait before considering that a response has not occurred, it is 14 μs.

5.5 Real-Time Scheduling

A real-time scheduler schedules real-time tasks sharing a resource. The goal of the real-time scheduler is to ensure that the timing constraints of these tasks are satisfied. The scheduler decides, based on the task timing constraints, which task to execute or to use the resource at any given time.

Traditionally, real-time schedulers are divided into offline and online schedulers. Offline schedulers make all scheduling decisions before the system is executed. At run-time a simple dispatcher is used to activate tasks according to the schedule generated before run-time. Online schedulers, on the other hand,

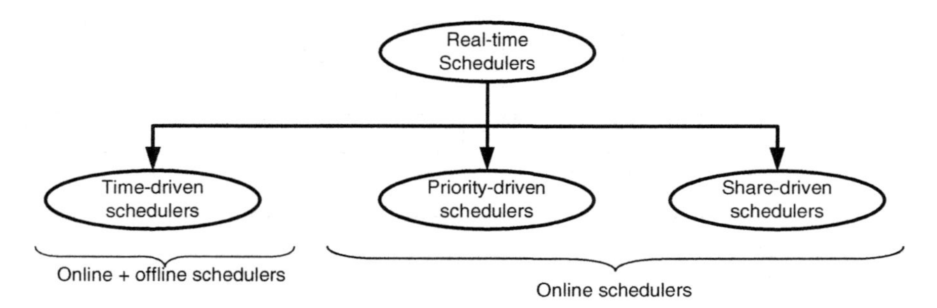

Fig. 5.6 Real-time schedulers

make scheduling decisions based on the system's timing constraints during run-time.

There are different schedulers developed in the research community [24, 25], only the basic concepts of different types of schedulers are presented here.

The schedulers are divided into three categories: time-driven schedulers, priority-driven schedulers and share-driven schedulers. This classification of real-time schedulers is depicted in Fig. 5.6.

Note that there also exist combinations of the predicted time-driven schedulers and the more flexible priority-driven schedulers and there exists methods to convert one policy to another.

5.5.1 Time-Driven Scheduling

Time-driven schedulers [26] work in the following way: The scheduler creates a schedule (exchange table). Usually the schedule is created before the system is started (offline), but it can also be done during run-time (online). At run-time, a dispatcher follows the schedule, and makes sure that tasks are executing at their predetermined time slots.

By creating a schedule offline, complex timing constraints, such as irregular task arrival patterns and precedence constraints, can be handled in a predictable manner that would be difficult to do online during run-time (tasks with precedence constraints require a special order of task executions, e.g., task A must execute before task B). The schedule that is created offline is the schedule that will be used at run-time. Therefore, the online behavior of time-driven schedulers is very predictable. Because of this predictability, time-driven schedulers are the more commonly used schedulers in applications that have very high safety-critical systems. However, since the schedule is created offline, the flexibility is very limited, in the sense that as soon as the system will change (due to adding of functionality or change of hardware), a new schedule has to be created and given to dispatcher. Creating a new schedule non-trivial and sometimes very time consuming. This motivates the usage of priority-driven schedulers described below.

5.5.2 *Priority-Driven Scheduling*

Scheduling policies that make their scheduling decisions during run-time are classified as online schedulers. These schedulers make their scheduling decisions online based on the system's timing constraints, such as, task priority. Schedulers that base their scheduling decisions on task priorities are called priority-driven schedulers.

Using priority-driven schedulers the flexibility is increased (compared to time-driven schedulers), since the schedule is created online based on the currently active task's properties. Hence, priority-driven schedulers can cope with change in work-load as well as adding and removing of tasks and functions, as long as the schedulability of the complete task-set is not violated. However, the exact behavior of the priority-driven schedulers is hard to predict. Therefore, these schedulers are not used as often in the most safety-critical applications.

Priority-driven scheduling policies can be divided into Fixed Priority Schedulers (FPS) and Dynamic Priority Schedulers (DPS). The difference between these scheduling policies is whether the priorities of the real-time tasks are fixed or if they can change during execution (i.e. dynamic priorities).

5.5.2.1 Fixed Priority Schedulers

When using FPS, once priorities are assigned to tasks they are not changed. Then, during execution, the task with the highest priority among all tasks that are available for execution is scheduled for execution. Priorities can be assigned in many ways, and depending on the system requirements some priority assignments are better than others. For instance, using a simple task model with strictly periodic non-interfering tasks with deadlines equal to the period of the task, a Rate Monotonic (RM) priority assignment has been shown by Liu and Layland [27] to be optimal in terms of schedulability. In RM, the priority is assigned based on the period of the task. The shorter the period is the higher priority will be assigned to the task.

5.5.2.2 Dynamic Priority Schedulers

The most well known DPS is the Earliest Deadline First (EDF) scheduling policy [27]. Using EDF, the task with the nearest (earliest) deadline among all tasks ready for execution gets the highest priority. Therefore the priority is not fixed, it changes dynamically over time. For simple task models, it has been shown that EDF is an optimal scheduler in terms of schedulability. Also, EDF allows for higher schedulability compared with FPS. Schedulability is in the simple scenario guaranteed as long as the total load in the scheduled system is $\leq 100\%$, whereas FPS in these simple cases has a schedulability bound of about 69%. For a good comparison between RM and EDF interested readers are referred to [24].

Other DPS are Least Laxity First (LLF) (sometimes also called Least Slack Time first (LST)) [28]. Here the priorities of the tasks are generated at the time of scheduling from the amount of laxity (for LLF, or slack for LST) available before the deadline is violated. Laxity (or slack time) is defined as the maximum time a task can be delayed on its activation and still complete within its deadline [24].

5.5.3 Share-Driven Scheduling

Another way of scheduling a resource is to allocate a share [29] of the resource to a user or task. This is useful, for example, when dealing with aperiodic tasks when their behavior is not completely known. Using share-driven scheduling it is possible to allocate a fraction of the resource to these aperiodic tasks, preventing them from interfering with other tasks that might be scheduled using time-driven or priority-driven scheduling techniques.

In order for the priority-driven schedulers to cope with aperiodic tasks, different service methods have been presented. The objective of these service methods is to give a good average response-time for aperiodic requests, while preserving the timing constraints of periodic and sporadic tasks. These services can be implemented as share-driven scheduling policies, either based on General Processor Sharing (GPS) [30, 31] algorithms, or using special server based schedulers [24, 32–34]. In the scheduling literature many types of servers are described, implementing server-based schedulers. In general, each server is characterised partly by its unique mechanism for assigning deadlines (for DPS based servers), and partly by a set of parameters used to configure the server. Examples of such parameters are priority (for FPS based servers), bandwidth, period, and capacity.

5.5.3.1 Share-Driven Scheduling in Fixed Priority Systems

Several server-based schedulers for FPS systems exist where the simplest one is the Polling Server (PS) [32]. A polling server allocates a share of the CPU to its users. This share is defined by the server's period and capacity, i.e., the PS is guaranteed to allow its users to execute within the server's capacity during each server period. The server is scheduled according to RM together with the normal tasks (if existing) in the system. However, a server never executes by itself. A server will only mediate the right to execute for its users, if some of its users have requested to use the server's capacity. Otherwise the server's capacity will be left unused for that server period. However, if the PS is activated and no user is ready to use the server capacity, the capacity is lost for that server period and the server's users have to wait to the next server period to be served. Hence, the worst-case service a user can get is when it requests capacity right after the server is activated (with its capacity replenished). The behavior of a PS server is in the worst-case equal to a task with the period of the server's period, and a worst-case execution time equal to the server's capacity. Hence, the analysis of a system running PS is straightforward.

Another server-based scheduler for FPS systems that is slightly better than the PS (in terms of temporal performance) is the Deferrable Server (DS) [34]. Here, the server is also implemented as a periodic task scheduled according to RM together with the (if existing) other periodic tasks. The difference from PS is that the server is not polling its users, i.e., checking if there are any pending users each server period and if not drop all its capacity. Instead, the DS preserves its capacity throughout the server period allowing its users to use the capacity at any time during the server period. As with the PS, the DS replenish its capacity at the beginning of each server period. In general, the DS is giving better response times than the PS. However, by allowing the servers' users to execute at any time during the servers' period it violates the rules govern by the traditional RM scheduling (where the highest priority task has to execute as soon it is scheduled), lowering the schedulability bound for the periodic task set. A trade-off to the DS allowing a higher schedulability but a slight degradation in the response times is the Priority Exchange (PE) algorithm. Here the servers' capacities are preserved by exchanging it for the execution time of a lower priority periodic task. Hence, the servers' capacities are not lost but preserved at the priority of the low priority task involved in the exchange. Note that the PE mechanisms are computationally more complex than the DS mechanisms, which should be taken into consideration in the trade-off.

By changing the way capacity is replenished, the Sporadic Server (SS) [32] is a server-based scheduler for FPS systems that allows high schedulability without degradation. Instead of replenishing capacity at the beginning of the server period, SS replenishes its capacity once the capacity has been consumed by its users. As DS, SS violates the traditional RM scheduling by not executing the highest priority task once it is scheduled for execution. However, this violation does not impact on the schedulability as the same schedulability bound is offered for a system running both with and without SS.

There are server-based schedulers for FPS systems having better performance in terms of response-time. However, this usually comes at a cost of high computational and implementation complexity as well as high memory requirements. One of these schedulers is the Slack Stealer. It should be noted that there are no optimal algorithms in terms of minimising the response time. The non existence of an algorithm that can both minimise the response time offered to users and at the same time guarantees the schedulability of the periodic tasks has been proven in [35]. Hence, there is a trade-off between response-time and schedulability when finding a suitable server-based scheduler for the intended target system.

5.5.3.2 Share-driven Scheduling in Dynamic Priority Systems

Looking at DPS systems, a number of server-based schedulers have been developed over the years. Many of the server-based schedulers for FPS systems have also been extended to EDF based DPS systems, e.g., an extension of PE called the Dynamic Priority Exchange (DPE) [36], and an extension of the SS called the Dynamic Sporadic Server (DSS) [36]. A very simple (implementation wise) server-based scheduler that provides faster response-times compared with SS yet not violating the overall load of the system (causing other tasks to miss their deadlines) is the Total Bandwidth Server (TBS) [36]. TBS makes sure that the server never uses more bandwidth than allocated to it (under the assumption that the users do not consume more capacity than specified by their worst-case execution times), yet providing a fast response time to its users (i.e., assigning its users with a close deadline as the system is scheduled according to EDF). Also, TBS has been enhanced by improving its deadline assignment rule [36]. A quite complex server-based scheduler is the Earliest Deadline Late server (EDL) [36] (which is a DPS version of the Slack Stealer). Moreover, there is an Improved Priority Exchange (IPE) [36] which has similar performance as the EDL, yet being less complex implementation wise. When the worst-case execution times are unknown, the Constant Bandwidth Server (CBS) [24] can be used, guaranteeing that the server's users will never use more than the server's capacity.

5.6 Real-Time Analysis

Time-driven schedulers create a schedule offline. As the schedule is created, the schedule is verified so that all timing constraints are met. However, both priority-driven and share-driven schedulers have a more dynamic behavior since the scheduling is performed during run-time. Here, timing analysis (schedulability tests) can be used in order to determine whether the temporal performance of a real-time system can be guaranteed for a certain task set scheduled by a certain scheduler. If such a guarantee is possible, the task set is said to be feasible.

There exist three different approaches for pre-run-time schedulability analysis: utilisation-based tests, demand-based tests and response-time tests. The first approach is based on the utilisation of the task-set under analysis (utilisation-based tests), the second is based on the processor demand at a given time interval (demand-based tests), and the third approach is based on calculating the worst-case response-time for each task in the task-set (response-time tests). Utilisation-based tests are usually less complex and faster to perform compared with demand-based tests and response-time tests, but they can not always be used for complicated task models [37].

5.6.1 Task Model

The task model notation used throughout this section is presented in Table 5.2.

Periodic tasks could be of two types: synchronous periodic tasks, and asynchronous periodic tasks. These are defined as:

Table 5.2 Task model notation

Abbrivation	Description
N	Number of tasks in the task set
C	Worst-case Execution time (WCET)
T	Period
r	Release time
D	Relative deadline
d	Absolute deadline
B	Blocking-time
R	Response-time
i	Task under analysis
hp(i)	set of tasks with priority higher than that of tasks i
lep(i)	set of tasks with priority less than or equal to that of tasks i

Synchronous periodic tasks are a set of periodic tasks where all first instances are released at the same time, usually considered time zero.

Asynchronous periodic tasks are a set of periodic tasks where tasks can have their first instances released at different times.

5.6.2 Utilisation-Based Tests

Seminal work on utilisation-based tests for both fixed priority schedulers and dynamic priority schedulers have been presented by Liu and Layland [27].

5.6.2.1 Fixed Priority Schedulers

In [27] by Liu and Layland, a utilisation-based test for synchronous periodic tasks using the Rate Monotonic (RM) priority assignment is presented (Liu and Layland provided the formal proofs). The task model they use consists of independent periodic tasks with deadline equal to their periods. Moreover, all tasks are released at the beginning of their period and have a known worst-case execution time and they are fully pre-emptive. If the test succeeds, the tasks will always meet their deadlines given that all the assumptions hold. The test is as follows:

$$\sum_{i=1}^{N} \frac{C_i}{T_i} \leq N \times \left(2^{1/N} - 1\right) \tag{5.1}$$

This test only guarantees that a task-set will not violate its deadlines if it passes this test. The lower bound given by this test is around 69% when N approaches infinity. However, there are task-sets that may not pass the test, yet they will meet all their deadlines. Later on, Lehoczky showed that the average case real feasible utilization is about 88% when using random generated task sets. Moreover, Lehoczky also developed an exact analysis. However, the test developed by Lehoczky is a much more complex inequality compared to Inequality (5.1). It has also been shown that, by having the task's periods harmonic (or near harmonic), the schedulability bound is up to 100% [38]. Harmonic task sets have only task periods that are multiples if each other.

Inequality (5.1) has been extended in various ways, e.g., by Sha et al. [39] to also cover blocking-time, i.e., to cover for when higher priority tasks are blocked by lower priority tasks. For a good overview of FPS utilisation-based tests interested readers are referred to [25].

5.6.2.2 Dynamic Priority Schedulers

Liu and Layland [27] also present a utilisation-based test for EDF (with the same assumptions as for Inequality (5.1)):

$$\sum_{i=1}^{N} \frac{C_i}{T_i} \leq 1 \tag{5.2}$$

This inequality is a necessary and sufficient condition for the task-set to be schedulable. It has been shown that the Inequality (5.2) is also valid for asynchronous task sets. However, later it has been shown that it is enough to investigate synchronous task sets in order to determine if a periodic task set is feasible or not [40].

5.6.3 Demand-Based Tests

The processor demand is a measure that indicates how much computation that is requested by the system's task set, with respect to timing constraints, in an arbitrary time interval $t \in [t_1, t_2)$. The processor demand $h_{[t1,t2)}$ is given by

$$h_{[t_1,t_2)} = \sum_{t_1 \leq r_k, d_k \leq t_2} C_k \tag{5.3}$$

where r_k is the release time of task k and d_k is the absolute deadline of task k, i.e., the processor demand is in an arbitrary time interval given by the tasks released within (and having absolute deadlines within) this time interval.

Looking at synchronous task sets, (5.3) can be expressed as $h(t)$ given by

$$h(t) = \sum_{D_i \leq t} \left(\left[1 + \left\lfloor \frac{t - D_i}{T_i} \right\rfloor \right] \times C_i \right) \tag{5.4}$$

where D_i is the relative deadline of task i. Then, a task set is feasible iff

$$\forall t, h(t) \leq t \tag{5.5}$$

for which several approaches have been presented determining a valid (sufficient) t [40, 41].

5.6.3.1 Dynamic Priority Schedulers

By looking at the processor demand, Inequality (5.2) has been extended for deadlines longer than the period [40]. Moreover, [41] present a processor demand-based feasibility test that allows for deadlines shorter than period. Given that Inequality (5.2) is satisfied, Spuri et al. [33] introduce a generalized processor demand-based feasibility test that allows for non pre-emptive EDF scheduling. Additional extensions covering sporadic tasks is discussed in [40].

5.6.4 Response-Time Tests

Response-time tests are calculating the behavior of the worst-case scenario that can happen for any given task, scheduled by a specific real time scheduler. This worst case behavior is used in order to determine the worst-case response-time for that task.

5.6.4.1 Fixed Priority Schedulers

Joseph and Pandya presented the first response-time test for real-time systems [42]. They present a response-time test for pre-emptive fixed-priority systems. The worst-case response-time is calculated as follows:

$$R_i = I_i + C_i \tag{5.6}$$

where I_i is the interference from higher priority tasks defined as:

$$I_i = \sum_{j \in hp(i)} \left(\left\lceil \frac{R_i}{T_j} \right\rceil \times C_j \right) \tag{5.7}$$

where $hp(i)$ is the set of tasks with higher priority than task i.

For FPS scheduled systems, the critical instant is given by releasing task i with all other higher priority tasks at the same time, i.e., the critical instant is generated when using a synchronous task set [32]. Hence, the worst-case response-time for task i is found when all tasks are released simultaneously at time 0.

The worst-case response-time is found when investigating the processors level-i busy period, which is defined as the period preceding the completion of task i, i.e., the time in which task i and all other higher priority tasks still not yet have executed until completion. Hence, the processors level-i busy period is given by rewriting (5.6) to:

$$R_i^{n+1} = \sum_{j \in hp(i)} \left(\left\lceil \frac{R_i^n}{T_j} \right\rceil \times C_j \right) + C_i \tag{5.8}$$

Note that (5.8) is a recurrence relation, where the approximation to the $(n+1)$th value is found in terms of the nth approximation. The first approximation is set to $R_i^0 = C_i$. A solution is reached when $R_i^{n+1} = R_i^n$, i.e., a so called fixed-point iteration. The recurrence equation will terminate given that Inequality (5.2) is fulfilled [32].

The work of Joseph and Pandya [42] has been extended by Audsley et al. [43] to cover for the non pre-emptive fixed-priority context. Note that non preemption introduces a blocking factor B_i due to execution already initiated by lower priority tasks. As a lower priority task has started its execution, it can not be pre-empted; hence, it might block higher priority tasks for the duration of its worst-case execution time. Also, in a non pre-emptive system, the processors level-i busy period is not including the task i itself:

$$R_i^{n+1} = B_i + \sum_{j \in hp(i)} \left(\left\lceil \left\lfloor \frac{R_i^n - C_i}{T_j} \right\rfloor + 1 \right\rceil \times C_j \right) + C_i \tag{5.9}$$

where the blocking factor B_i is defined as follows:

$$B_i = \begin{cases} 0 & \text{if} \quad lep(i) = 0 \\ \max\limits_{k \in lep(i)} \{\mathrm{C_k} - \varepsilon\} & \text{if} \quad lep(i) \neq 0 \end{cases} \tag{5.10}$$

where $lep(i)$ is the set of tasks with priority less or equal than task i. ε is the minimum time quantum, which, in computer systems, corresponds to one clock cycle. Including ε makes the blocking expression less pessimistic by safely removing one clock cycle from the worst case execution time of task k, causing the blocking. The recurrence equation (5.9) is solved in the same way as (5.8). Note that the in the presence of blocking, the scenario giving the critical instant is re-defined. The maximum interference now occurs when task i and all other higher priority tasks are simultaneously released just after the release of the longest lower priority task (other than task i).

5.6.4.2 Dynamic Priority Schedulers

In dynamic priority systems, the worst-case response-time for a task-set is not necessarily obtained considering the processors level-i busy period generated by releasing all tasks at time 0. Instead, Spuri [33] find the worst-case response-time in the processors deadline-i busy period. The deadline-i busy period is the period

Table 5.3 Comparison of networks

Serial/Field bus features	CAN	TTCAN	TOKEN BUS	Ethernet (CSMA/CD)	MIL-STD- 1553B
Standard	CAN data link layer: ISO 11898-1 CAN Physical link layer: ISO11898-1	IEEE 802.4 network protocol ISO 11898-4	Bus control network for logically arranged nodes	CSMA/CD MAC protocol: IEEE 802.3 network protocol	MILITARY STANDARD
Communication scheduling	Event base, message based arbitration on message priority	Event base, Exclusive time window for a message only	Event based on token passing	Event based	Command response; activity on the bus initiated by BC, RT has to respond to BC
Maximum data rate	1 MB/S up to 40 m distance	1 MB/S up to 40 m distance	5 MB/S up to 1000 m	10 MB/S up to 2500 m	1 MB/S, NO limit
Maximum data size with overheads	8 bytes Overhead:47/65 bits	8 bytes Overhead:47/ 65 bits	32000 bytes Overhead:80 bits minimum	1500 bytes Overhead:206 bits minimum	32 data word (512 actual data bits) overhead = 208 bits
Data flow	Half duplex	Half duplex	Half duplex	Half duplex / Full duplex	Half duplex
MAC layer	CSMA /CD/ AMP Arbitration on message priority, Priority based	Exclusive window for one message, Arbitration for rest of messages like basic CAN	Token passing	CSMA/CD with BEB algorithm	TDMA, Command response
Clock synchronization	Not required	Reference message is used to for local synchronization	Not required	Not required	Not required
Delay and jitter	Low priority messages has higher delay jitter	low delay and jitter improvement	Worst case delay is fixed and low jitter	Unbounded delay and very high jitter	Worst case delay is fix with very low jitter
Network size in term of node	120 nodes	120 nodes	100 nodes	Maximum segment length is 100 m. Minimum length between nodes is 2.5 m. Maximum number of connected segments is 1024. Maximum number of nodes per segment is 1 (star topology).	31 remote terminal maximum

(continued)

Table 5.3 (continued)

Serial/Field bus features	CAN	TTCAN	TOKEN BUS	Ethernet (CSMA/CD)	MIL-STD- 1553B
Physical Media	Twisted pair Single wire CAN (CSMA/CR)	Twisted pair	Twisted pair	Twisted pair	Twisted pair
Topology	Multidrop	Multidrop	Logical ring	10BASE5 uses bus topology	Multidrop
Redundancy	NO	NO	NO	NO	DUAL or more redundant BUS
Fault Tolerant	NO	NO	NO	NO	YES
Fail silent	YES, Node can go into BUS OFF state	YES, Node can go into BUS OFF state	NO	NO	NO
Frame/Message check	YES, acknowledgement bit, CRC Incremental/ Detrimental counter for node state on the bus	YES, acknowledgement bit, CRC Incremental/Detrimental counter for node state on the bus	CRC	CRC	Parity, status word
Node failure tolerance	YES	YES	YES	YES	Yes, if the BUS controller remains functional

in which only tasks with absolute deadlines smaller than or equal to d_i are allowed to execute.

Hence, in dynamic-priority systems the worst-case response-time for an arbitrary task i can be found for the pre-emptive case when all tasks, but i, are released at time 0. Then, multiple scenarios have to be examined where task i is released at some time t.

Also for the non pre-emptive case, all tasks but i are released at time 0. However, one task with an absolute deadline greater than task i (i.e., one lower priority task) has initiated its execution at time $0-\varepsilon$. Then, as in the preemptive case, multiple scenarios have to be examined where task i is released at some time t.

Worst-case response-time equations for both preeptive and non pre-emptive EDF scheduling are given by Spuri [33]. Furthermore, these have been extended for response-time analysis of EDF scheduled systems to include offsets.

5.7 Comparison of Networks

Comparisons of some networks are listed in Table 5.3.

5.8 Summary

In this chapter, MAC mechanism of three candidate networks is presented in detail. The MAC mechanism is responsible for the access to the network medium and hence affects the timing requirement of message transmission. Comparison of network parameters is also presented. These comparisons provide an understanding of these network protocols, and can be used as a primary guideline for selecting a network solution for a given application.

References

1. Tanenbaum A (2003) Computer networks. Prentice Hall, Upper Saddle River
2. Nolte T (2006) Share-driven scheduling of embedded networks. PhD thesis, Malardalen University, Sweden, May 2006
3. IEEE 802.15, Working group for wireless personal area networks (wpans), http://www.ieee802.org/15/
4. MIL-STD-1553B: Aircraft internal time division command/response multiplex data bus, 30 April 1975
5. Kopetz H, Bauer G (2003) The time-triggered architecture. Proc IEEE 91(1):112–126
6. Berwanger J, Peller M, Griessbach R. Byteflight—a new high-performance data bus system for safety-related applications. BMW AG, London

7. Flexray communications system—protocol specification, version 2.0, 2004
8. Malcolm M, Zhao W (1994) The timed token protocol for real-time communication. IEEE Comput 27(1):35–41
9. IEC 61158: Digital data communications for measurement and control: Fieldbus for use in industrial control systems, 2003
10. PROFInet - architecture description and specification, No. 2.202, 2003
11. Specification of the ttp/a protocol, 2005, http://www.ieee802.org/15/
12. Spurgeon CE (2000) Ethernet: the definitive guide. O'Reilly & Associates, Inc, USA
13. Shoch JF, Dalal YK, Redell DD, Crane RC (1982) Evolution of the ethernet local computer network. Computer 15(8):10–27
14. Kopetz H, Damm A, Koza C, Mulazzani M, Schwabl W, Senft C, Zainlinger R (1989) Distributed fault-tolerant real-time systems: The mars approach. IEEE Micro 9(1):25–40
15. Chiueh T, Venkatramani C (1994) Supporting real-time traffic on ethernet. In: Proceedings of Real-Time Systems Symposium, pp 282–286
16. Pedreiras P, Almeida L, Gai P (2002) The ftt-ethernet protocol: Merging flexibility, timeliness and efficiency. In: Proceedings of the 14th Euromicro Conference on Real-Time Systems, 0:152
17. Molle M, Kleinrock L (1985) Virtual time CSMA: why two clocks are better than one. IEEE Trans Commun 33(9):919–933
18. Zhao W, Stonkovic JA, Ramamritham K (1990) A window protocol for transmission of time-constrained messages. IEEE Trans Comput 39(9):1186–1203
19. Kweon S-K, Shin KG, Workman G (2000) Achieving real-time communication over ethernet with adaptive traffic smoothing. In: Real-Time and Embedded Technology and Applications Symposium, IEEE, 0:90
20. Lo Bello L, Kaczynski GA, Mirabella O (2005) Improving the real-time behavior of ethernet networks using traffic smoothing. IEEE Trans Ind Inform 1(3):151–161
21. Farsi M, Ratcliff K, Barbosa M (1999) An overview of controller area network. Comput Control Eng J 10:113–120
22. CAN specification 2.0. part A and B, CAN in automation (CiA)
23. MIL-STD-1553 Tutorial, CONDOR Engineering, Inc, Santa Barbara, CA 93101
24. Buttazzo GC (2003) Hard real-time computing systems - predictable scheduling algorithms and applications. Springer, Heidelberg
25. Sha L, Abdelzaher T, Arzen K-E, Cervin A, Baker T, Burns A, Buttazzo G, Caccamo M, Lehoczky J, Mok AK (2004) Real time scheduling theory: A historical perspective. Real-Time Syst 28(2–3):101–155
26. Kopetz H (1998) The time-triggered model of computation. In: Proceedings of the 19th IEEE Real-Time Systems Symposium (RTSS'98), pp 168–177
27. Liu CL, Layland JW (1973) Scheduling algorithms for multiprogramming in a hard real-time environment. J ACM 20(1):40–61
28. Leung JY-T, Whitehead J (1982) On the complexity of fixed priority scheduling of periodic real-time tasks. Perform Eval 2(4):237–250
29. Stocia I, Abdel-Wahab H, Jeffay K, Baruah SK, Gehrke JE, Plaxton CG (1996) A proportional share resource allocation algorithm for real-time, time-shared systems. In: Proceedings of the 17th IEEE Real-Time Systems Symposium (RTSS'96), pp 288–299
30. Parekh AK, Gallager RG (1993) A generalized processor sharing approach to flow control in integrated services networks: the single node case. IEEE/ACM Trans Netw 1(3):334–357
31. Parekh AK, Gallager RG (1994) A generalized processor sharing approach to flow control in integrated services networks: the multiple node case. IEEE/ACM Trans Netw 2(2):137–150
32. Sprunt B, Sha L, Lehoczky JP (1989) Aperiodic task scheduling for hard real-time systems. Real-Time Syst 1(1):27–60
33. Spuri M, Buttazzo GC (1994) Efficient aperiodic service under earliest deadline scheduling. In: Proceedings of the 15th IEEE Real-Time Systems Symposium (RTSS'94), pp 2–11
34. Strosnider JK, Lehoczky JP, Sha L (1995) The deferrable server algorithm for enhanced aperiodic responsiveness in the hard real-time environment. IEEE Trans Comput 44(1):73–91

35. Tia T-S, Liu W-S, Shankar M (1996) Algorithms and optimality of scheduling soft aperiodic requests in fixed-priority preemptive systems. Real-Time Syst 10(1):23–43
36. Spuri M, Buttazzo GC (1996) Scheduling aperiodic tasks in dynamic priority systems. Real-Time Syst 10(2):179–210
37. Tindell KW, Burns A, Wellings AJ (1994) An extendible approach for analysing fixed priority hard real-time tasks. Real-Time Syst 6(2):133–151
38. Sha L, Goodenough JB (1990) Real-time scheduling theory and ADA. IEEE Comput 23(4):53–62
39. Sha L, Rajkumar R, Lehoczky JP (1990) Priority inheritance protocols: An approach to real-time synchronization. IEEE Trans Comput 39(9):1175–1185
40. Baruah SK, Mok AK, Rosier LE (1990) Preemptive scheduling hard real-time sporadic tasks on one processor. In: Proceedings of the 11th IEEE Real-Time Systems Symposium (RTSS'90), pp 182–190
41. Baruah SK, Rosier LE, Howell RR (1990) Algorithms and complexity concerning the preemptive scheduling of periodic real-time tasks on one processor. Real-Time Syst 2(4):301–324
42. Joseph M, Pandya P (1986) Finding response times in a real-time system. Comput J 29(5):390–395
43. Audsley NC, Burns A, Richardson MF, Tindell K, Wellings AJ (1993) Applying new scheduling theory to static priority pre-emptive scheduling. Softw Eng J 8(5):284–292

Chapter 6
Response-Time Models and Timeliness Hazard Rate

6.1 Introduction

For networked systems, modeling of delay or response-time distribution [1–6] plays an important role. It helps in estimating the probability of missing a specified deadline [7], analyzing effect of redundancies on response-time. For dependability models considering timeliness failures, estimation of timeliness hazard rate is required. In this chapter, probabilistic response-time models for CAN, MIL-STD-1553B and Ethernet networks is derived, effect of redundancies on response-time of these networks is analyzed, a method to estimate timeliness hazard rate is proposed.

6.2 Review of Response-Time Models

6.2.1 Tagged Customer Approach

In tagged customer approach, an arbitrary message/customer is picked as the tagged message/customer and its passage through the network (closed queuing, with finite steady-state distribution) is tracked. By this method, the problem of computing the response time distribution of the tagged customer is transformed into time to absorption distribution of a finite-state, continuous time Markov chain (CTMC), conditioned on the state of the system upon entry. Using the arrival theorem of Sevcik and Mitrani [8], distribution of the other customers in the network at the instant of arrival of tagged customer can be established. This allows obtaining the unconditional response time distribution.

A. K. Verma et al., *Dependability of Networked Computer-based Systems*,
Springer Series in Reliability Engineering, DOI: 10.1007/978-0-85729-318-3_6,

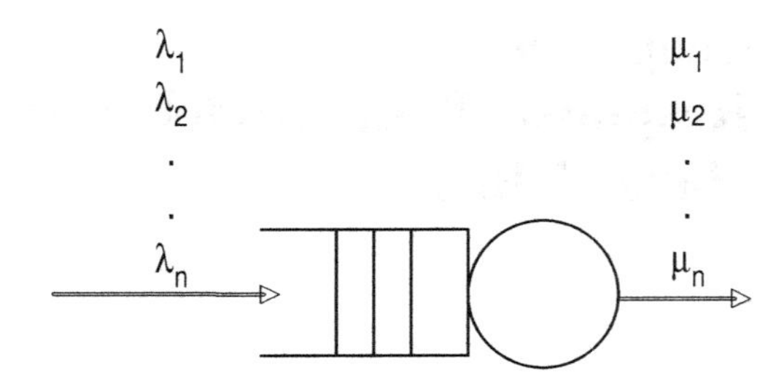

Fig. 6.1 Single server queuing model

Theorem:

$$\lim_{k\to\infty} P[Y_{k-n}] = \pi_{N-1}(n)$$

where

$$n \in S(N-1)$$

This theorem forms the basis of the "tagged customer" approach for computing the response time distribution. It is also referred to as the **Arrival Theorem**, and states that in a closed queueing network an arriving customer would see the network in equilibrium with one less customer. Thus in a network with N customers, the tagged customer sees the network in equilibrium with $N-1$ customers. The arrival theorem gives the probability distribution for the state of the system as seen by the arriving customer. So, computing the response time distribution using tagged customer approach is a two step process [5]:

1. Compute the steady-state probabilities for each of the states of the queueing network with one less customer, $\pi_{N-1}(n)$
2. Use these probabilities to compute the response time distribution, $P[R \leq t]$

6.2.1.1 Example 1: A Single Server System

Let us consider a single server queueing model of a computing system as shown in Fig. 6.1. Here jobs after service leave the system, at the same time an identical customer becomes active and joins the queue following job dependent Poisson arrivals. We assume SPNP service discipline at the queue and service time distributions are job dependent exponential distribution. In the figure, number in the subscript denotes the priority of the job, lesser the number higher is the priority.

The system has 2 jobs. Fig. 6.2 shows the CTMC of the system. State (1.1) indicates availability of both jobs for service. Similarly, state (1,0) indicates availability of job '1' for service. λ_1 and λ_2 indicated the job arrival rate and μ_1 and μ_2 indicate the service rate of job '1' and job '2', respectively. For this model we define response time as the amount of time elapsed from the time instant at which job enter the queue until the instant at which it leaves the system after receiving service.

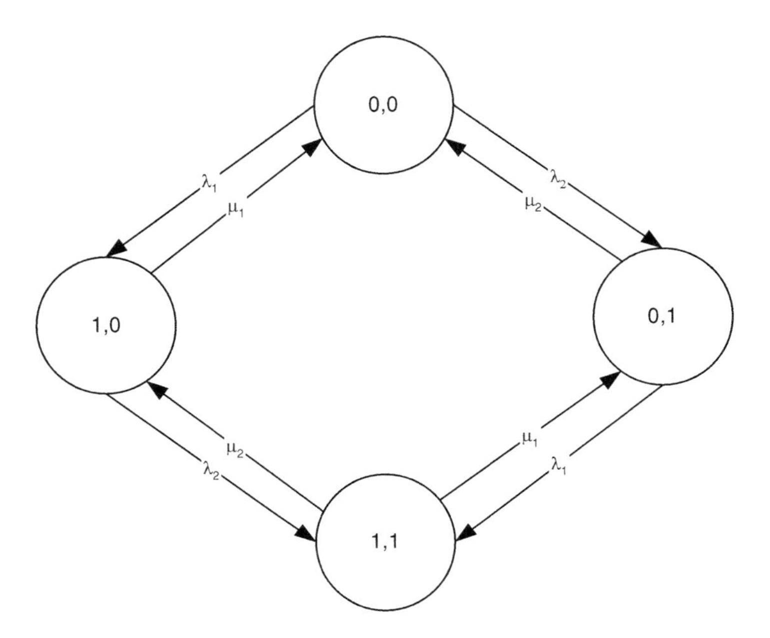

Fig. 6.2 Single server queuing model

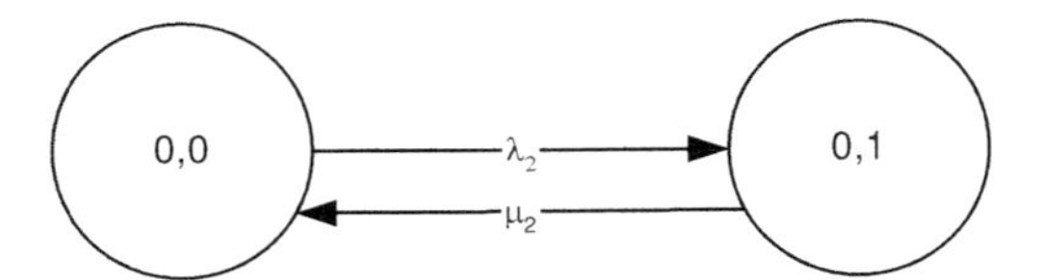

Fig. 6.3 CTMC without tagged customer

The first step is to derive CTMC of the system without the job of interest and finding the steady-state probabilities. The CTMC is shown in Fig. 6.3.

Having found the steady state probability of the system without tagged customer, we now construct the modified Markov chain from these state to state where tagged customer leaves the system. The modified CTMC is shown in Fig. 6.4.

In the figure stating and absorbing states are obvious. Following set of differential equations need to be solved to get passage times:

$$
\begin{aligned}
\frac{d(P_1)}{dx} &= -0.175P_1 + 0.025P_2 \\
\frac{d(P_2)}{dx} &= 0.125P_1 - 0.075P_2 \\
\frac{d(P_3)}{dx} &= 0.050P_1 - 0.125P_3 \\
\frac{d(P_4)}{dx} &= 0.050P_2 + 0.125P_3 - 0.225P_4 \\
\frac{d(P_5)}{dx} &= 0.025P_4 - 0.2P_5 \\
\frac{d(P_6)}{dx} &= 0.2P_4 \\
\frac{d(P_7)}{dx} &= 0.2P_5
\end{aligned}
$$

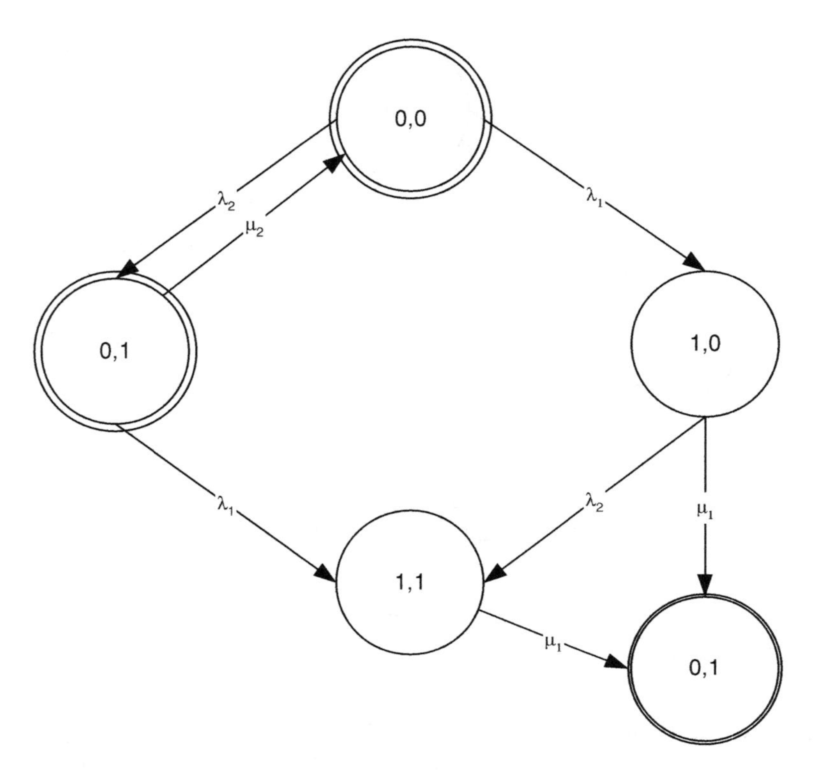

Fig. 6.4 Absorbing CTMC for response time of tagged customer

For following initial conditions:

1. $P_1(0) = 1$, for other states 0
2. $P_2(0) = 1$, for other states 0

with boundary condition: $\sum_{i=1}^{7} P_i = 1$

Finally both the passage times are combined using the steady-state probabilities calculated in the first step. So, the final result is:

$$P[R \leq t] = 0.1667\left(1 - \frac{8}{3}e^{\frac{-t}{8}} + \frac{5}{3}e^{\frac{-t}{5}}\right) + 0.8333\left(1 - e^{\frac{-t}{5}}\right)$$

This analytical response time is plotted along with simulation results in Fig. 6.5. From figure it is clear that response time distribution obtained using the proposed approach and from the simulation are quiet close. It proves the effectiveness of the proposed method.

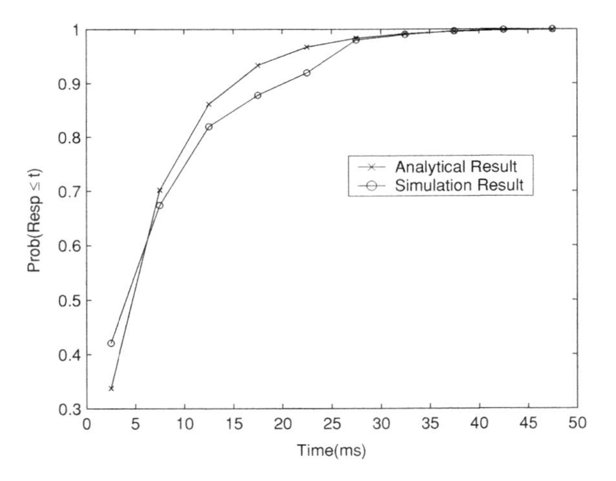

Fig. 6.5 Single server response—analytical and simulation

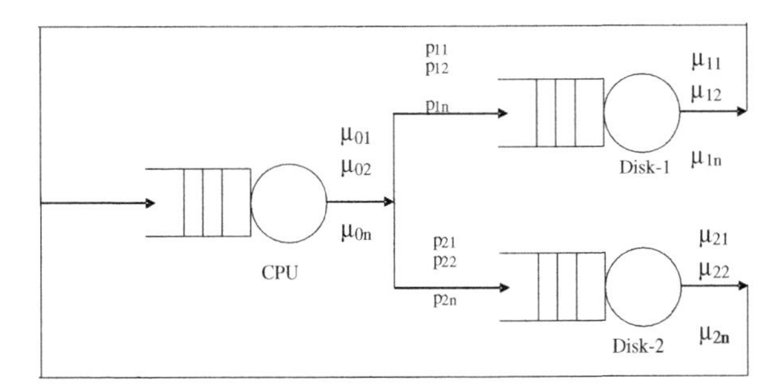

Fig. 6.6 Single server queuing model

6.2.1.2 Example 2: A Multi Server System

Now, let us take a multi-server model of a computer system as shown in Fig. 6.6. It is also an example of close-queueing network. We have already assumed that service discipline at all queues is SPNP and the service time distributions are job dependent exponential. The service rates for a job i at CPU, Disk-1 and Disk-2 are μ_{0i}, μ_{1i}, and μ_{2i}, respectively. When the customer finishes at the CPU, it will either access to Disk-1 or Disk-2 with probability p_{1i}, p_{2i}, respectively. After completing the disk access, the job rejoins the CPU queue. For this model we define the response time as the amount of time elapsed from the instant at which the job enters the CPU queue for service until the instance it leaves either of the disk.

Following the steps mentioned in section 6.2.1, CTMC of the system with job '1' in the system is prepared. The steady state probabilities are evaluated for job '2' being in CPU, Disk-1 and Disk-2 . These are the probabilities, job '1' may see at its arrival.

Now, CTMC is evolved from these states and sets of differential equations are made. These equations are solved for three different initial conditions, i.e. at arrival, job '1' finds job '2' at (1) CPU.

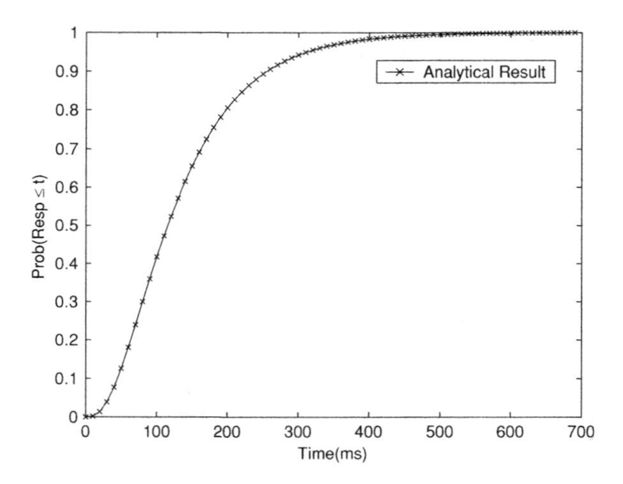

Fig. 6.7 Multi server response—analytical response

Table 6.1 Example system

Job	λ_i	P_i	PDF
T_1	0	5	U[2, 5]
T_2	3	10	U[1, 2]
T_3	6	15	U[1, 3]

After this, these three conditional passage times are unconditioned using Arrival Theorem and steady-state probabilities.

The response time distribution of the highest priority job is plotted in the Fig. 6.7.

6.2.2 Probabilistic Response-Time Model

The method illustrated is applicable if the model has following properties [2, 3]:

1. System uses preemptive priority driven scheduling policy
2. Tasks have fixed priority and computation time distribution follows uniform random distribution

To calculate the response time distribution of a job, this method takes into account not only the computation times required by the job and the interference that future jobs could cause on it due to preemption, but also the pending workload not yet serviced at the instant the job is released.

This method is based on determining pending workload, and convolution of the job's computation time with higher priority tasks computation times.

To illustrate the method, we have taken a preliminary example from Refs. [2, 3]. The system is modeled as a set of jobs $\{T_i\}$, each job being a three-tuple i, P_i, C_i where i is the release instant of the job, P_i is the priority under which the

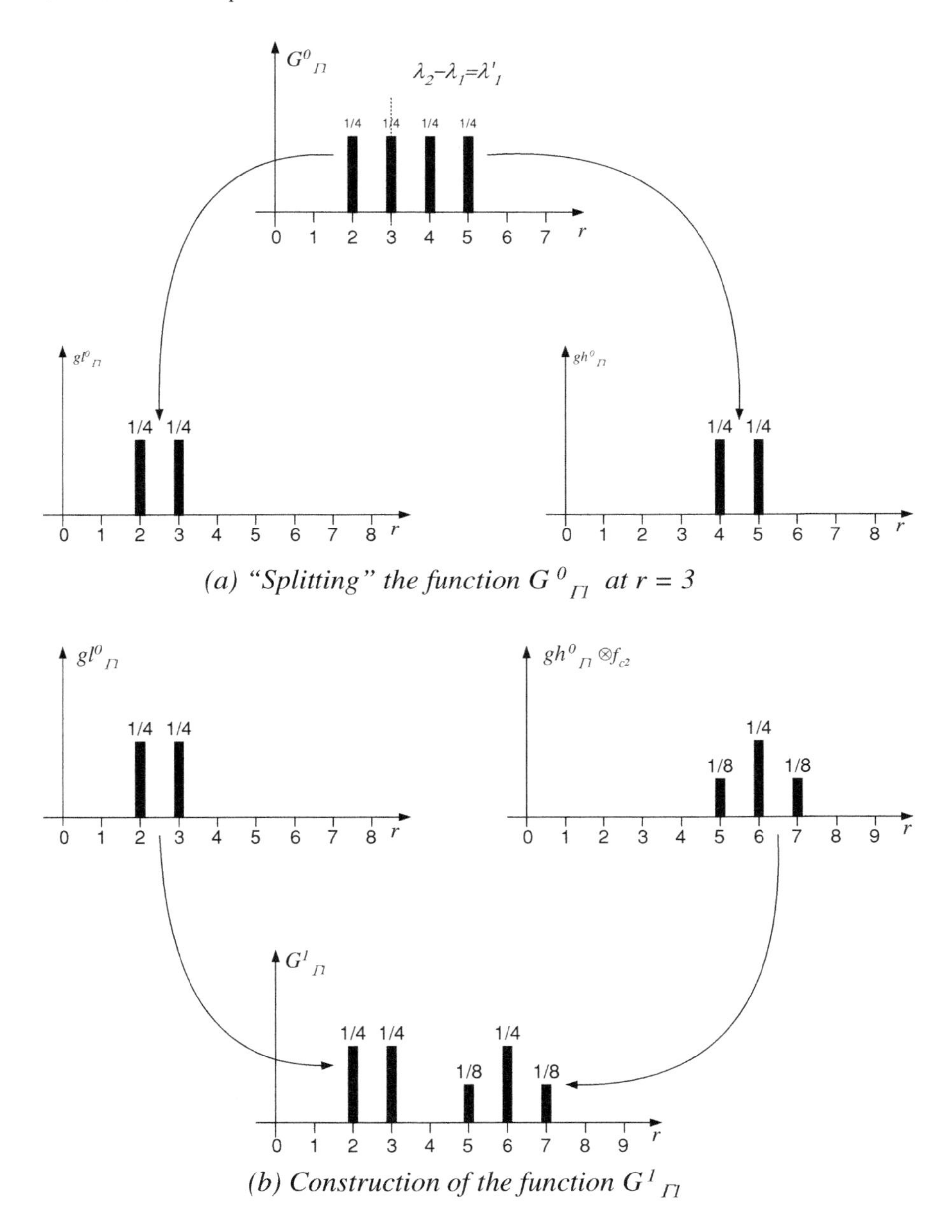

Fig. 6.8 Probabilistic response time estimation—graphical method

job runs, and C_i is the required computation time, which is a random variable with known probability density function (PDF).

Consider the system shown in Table 6.1, and it is required to obtain the PDF of the response time for the job T_1.

The three steps used to determine the PDF of response-time graphically, are shown in Fig. 6.8. These three steps are used iteratively to determine the PDF of response-time in complex system involving several jobs.

It is an iterative method whose computational complexity is a function of number of jobs in the system and the maximum number of points defining the computation times.

6.3 Response-Time Models

6.3.1 CAN

In this section, a CAN model is discussed. For a benchmark problem it is compared with literature, the results are compared with results from simulation model and the basic CAN model is improved and results are compared again.

6.3.1.1 Worst-Case Delay Analysis

CAN network delay is also referred as response-time of CAN in literature. Tindell et al. [9, 10] present analysis to calculate the worst-case latencies of CAN messages. The analysis is based on the standard fixed priority response time analysis.

The worst-case response-time of message is the longest time between the queueing of a message and the time message reaches at destination nodes. In case of CAN, it is defined to be composed of two delays, (1) queueing delay, q_i, (2) transmission delay, C_i [9]. The queueing delay is the longest time that a message can be queued at a node and be delayed because of other higher- and lower- priority messages are being sent on the bus. The transmission delay is the actual time taken to send the message on the bus. Thus, worst-case response-time is defined as:

$$R_i = q_i + C_i \tag{6.1}$$

The queueing time, q_i is itself composed of two times, (1) longest time that any lower priority message can occupy the bus, B, (2) the longest time that all higher priority messages can be queued and occupy the bus before the message i is finally transmitted.

$$q_i = B_i + \sum_{j \in hp(i)} \left\lceil \frac{q_i + J_j + \tau_{\text{bit}}}{T_j} \right\rceil C_j \tag{6.2}$$

where J_i is the queueing jitter of the messages, i.e., the maximum variation in the queueing time relative to $T_i, hp(i)$ is the set of messages with priority higher than i, τ_{bit} (bit-time) caters for the difference in arbitration start times at the different nodes due to propagation delays and protocol tolerances. Equation 6.2 is

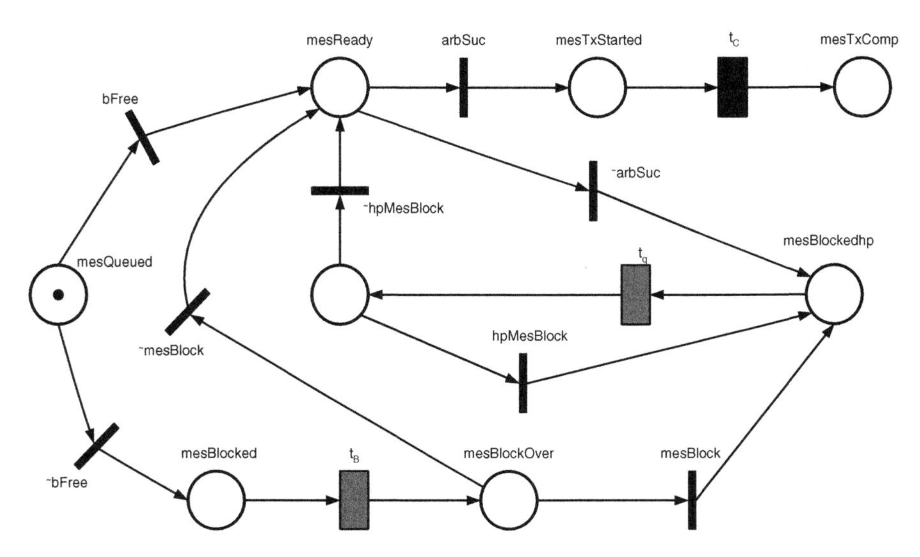

Fig. 6.9 Response-time model for CAN. A token in place *mesQueued* depicts that a message of interest is queued for transmission. Time taken by token to reach place *mesTxComp* is the response-time of message. Immediate transitions $bFree, hpmesBlock, arbSuc$ are probabilities of bus being free, blocking due to higher priority message, and successful arbitration, respectively. Immediate transitions labeled with as prefix $\sim$ are complementary to transitions without this symbol. General transitions t_B, t_q represent time associated with blocking, and queueing, respectively. Deterministic transition t_C, represent time associated with transmission

recurrence relation for q_i. Considering effect of external interference and error, the worst-case response time [11] can be given as:

$$R_i = B_i + \sum_{j \in hp(i)} \left\lceil \frac{q_i + J_j + \tau_{\text{bit}}}{T_j} \right\rceil C_j + C_i + E(q_i + C_i) \tag{6.3}$$

6.3.1.2 Basic CAN Model

The expression (6.1) depicts main parameters affecting response-time. The model presented here is based on following assumptions:

1. Queueing jitter, J_i is neglected.
2. Messages are not synchronized with each other.
3. Worst-case message transmission time is taken, i.e. with maximum number of stuff bits. (In the end, a method is discussed to accommodate bit-stuffing mechanism with the proposed model)
4. All parameters (*pdfs* and probabilities) are time invariant.

Response-time is defined as, the *time from the time instant a message is queued for transmission to completion of transmission*. Response-time model for a typical

CAN message is shown in Fig. 6.9. This model is based on Deterministic Stochastic Petri Net (DSPN). The model is analyzed analytically and the method is discussed in detail. This DSPN model is chosen for better representation and explanation of analysis steps. The DSPN model is only for explaining the model. A brief introduction to SPN, GSPN and DSPN is given in Chapter 3.

In Fig. 6.9 a token in place *mesQueued* depicts that a message of interest is queued for transmission. Time taken by token to reach place *mesTxComp* is the response-time of message. Immediate transitions *bFree, hpmesBlock, arbSuc* are probabilities of bus being free, blocking due to higher priority message, and successful arbitration, respectively. Immediate transitions labeled with as prefix $\sim$ are complementary to transitions without this symbol. General transitions t_B, t_q represent time associated with blocking and queueing, respectively. Deterministic transition t_C represent time associated with worst-case transmission.

To analyze this model, value of all transitions, immediate (probabilities) and timed (*pdfs*) are required. This model gives response-time distribution of one message only. So parameter values need to be calculated for all the messages whose response-time distribution is required.

Set of messages of the system are denoted by M. Parameter are estimated for a message m from set M. In parameters estimation sub-section, $i \in M$ means all messages except message m for which the parameter is being estimated.

Probability of finding bus free (*bFree*): probability that a message finds the network free when it gets queued, is estimated based on the utilization of network. This utilization is by other messages of network.

$$P^m_{\text{free}} = 1 - \sum_{i \in M} \frac{C_i}{T_i} \tag{6.4}$$

Probability of finding the bus free, by a Message m is the complement of utilization. This is because in a closed system (with fixed number of messages/ customers) with n messages, a message on arrival finds the system in equilibrium with $n-1$ messages [5].

Probability of no collision with high priority message (*arbSuc*): when network is free, a node with ready message can start transmission. Node will abort and back off transmission if it finds any higher priority message concurrently being transmitted. This can happen if a node start transmitting a higher priority message within the collision window τ_w.

$$P^m_{\text{Suc}} = \prod_{\substack{i \in M \\ i \in hp(m)}} P^i_C \tag{6.5}$$

where

$$\begin{aligned} P^i_C &= \text{Prob[non occurrence of ith message in time } \tau_w] \\ &= 1 - \left(\frac{1}{T_i} \times \tau_w\right) \end{aligned}$$

Blocking time (t_B): a message in queue can be blocked by any message under transmission by any of the other nodes. This is because CAN messages in transmission cannot be preempted. *pdf* of this blocking time $p_b(t)$ is obtained by following steps:

1. Find the ratio r_i of all the messages. $r_i = \frac{1}{T_i \sum_j \frac{1}{T_j}}$, for $i, j \in M$
2. Construct a *pdf*, $p(t)$ of total blocking time by other messages

$$p(t) = \sum_{i \in M} r_i \cdot \delta(t - C_i) \tag{6.6}$$

3. Message can get ready at any time during the blocking time with equal probability. So, effective blocking time is given by following convolution

$$p_b^m(t) = \frac{1}{\max(C_i)} \int_0^t p(\tau)[U(t + \tau) - U(t + \max(C_i) + \tau)]d\tau \tag{6.7}$$

Blocking time by high priority message (t_q): when the ready node finds bus free and start transmission of ready message, then if within the collision time window, another node starts transmitting a higher priority message, node backs off. And the message need to wait till the time of completion of this transmission. *pdf* of blocking time by high priority message, p_{bhp} is obtained by following steps 1–3 of *Blocking time* with one variation, instead of all messages only high priority message of network are considered.

1. Find the ratio r_i^H of all the messages. $r_i^H = \frac{1}{T_i \sum_j \frac{1}{T_j}}$, for $i, j \in M, i \in hp(m)$
2. *pdf* of blocking after back off is given by

$$p_{bhp}^m(t) = \sum_{i \in hp(m)} r_i^H \cdot \delta(t - C_i) \tag{6.8}$$

where $\delta(\cdot)$ is Dirac delta function.

Probability of no new higher priority message arrival in $t_B(\sim mesBlock)$: this is similar to *arbSuc* with the difference that instead of collision window time, mean of BlockTime is used.

$$P_{T_B}^m = \prod_{\substack{i \in M \\ i \in hp(m)}} P_{T_B}^i \tag{6.9}$$

where

$$\begin{aligned} P_{T_B}^i &= \text{Prob}[\text{non occurrence of ith message in time } BlockingTime] \\ &= 1 - \left(\frac{1}{T_i} \times E[t_B]\right) \end{aligned}$$

Probability of no new higher priority message arrival in $t_q(\sim hpMesBlock)$: this is similar to previous.

$$P^m_{T_{Bhp}} = \prod_{\substack{i \in M \\ i \in hp(m)}} P^i_{T_{Bhp}} \tag{6.10}$$

where

$$\begin{aligned} P^i_{T_{Bhp}} &= \text{Prob}[\text{non occurrence of ith message in time } \textit{Block Time by New}] \\ &= 1 - \left(\frac{1}{T_i} \times E[t_{Bhp}]\right) \end{aligned}$$

Queueing time: time to reach place *mesReady* from *mesBlockedhp* in *i*th step is modeled as a single r.v. with *pdf* $B^m_{hp}(i,t)$.

$$B^m_{hp}(i,t) = \left[\left(1 - P^m_{T_{Bhp}}\right)^{i-1} P^m_{T_{Bhp}}\right] \eta^i(t)$$

where

$$\begin{aligned} \eta^i(t) &= \eta^{i-1}(t) \otimes p^m_{bhp}(t) \\ \eta(t) &= p^m_{bhp}(t) \\ B^m_{hp}(t) &= \sum_i B_{hp^m}(i,t) \end{aligned}$$

Symbol $\otimes$ is used to denote convolution.

In the same way, time to reach place *mesTxStarted* in *i*th attempt from *mesReady* is modeled as a single r.v. with *pdf* $t_{rdy}(i,t)$.

$$t^m_{rdy}(i,t) = \left[(1 - P_{\text{free}})^{i-1} P_{\text{free}}\right] (B^m_{hp})^{i-1}(t)$$

where

$$\begin{aligned} (B^m_{hp})^i(t) &= (B^m_{hp})^{i-1}(t) \otimes B^m_{hp}(t) \\ (B^m_{hp})^0(t) &= \delta(t) \\ t^m_{rdy}(t) &= \sum_i t^m_{rdy}(i,t) \end{aligned}$$

From the instant message is queued, it can reach state *mesReady* either directly or via state *mesBlocked* and *mesBlockedhp*. State *mesBlocked* has an associated time delay. So, using total probability theorem [12], total queueing time is given as:

$$\begin{aligned} q^m(t) = {} & P^m_{\text{free}} t_{rdy}(t) + \left(1 - P^m_{\text{free}}\right) P^m_{T_B} \left[p^m_b(t) \otimes t_{rdy}(t)\right] \\ & + \left(1 - P^m_{\text{free}}\right)\left(1 - P^m_{T_B}\right)\left[p^m_b(t) \otimes p^m_{bhp}(t) \otimes t_{rdy}(t)\right] \end{aligned} \tag{6.11}$$

Table 6.2 SAE CAN messages used for analysis and comparison

Message ID	No. of bytes	Ti (ms)	Di (ms)
17	1	1000	5
16	2	5	5
15	1	5	5
14	2	5	5
13	1	5	5
12	2	5	5
11	6	10	10
10	1	10	10
9	2	10	10
8	2	10	10
7	1	100	100
6	4	100	100
5	1	100	100
4	1	100	100
3	3	1000	1000
2	1	1000	1000
1	1	1000	1000

Response-time: response-time of a message is sum of its queueing time and transmission time.

$$r^m(t) = q^m(t) \otimes C_m \delta(t) \tag{6.12}$$

Response-time distribution can be evaluated as

$$R^m(t) = \int_0^t r^m(\tau) d\tau \tag{6.13}$$

Let t_d^m is deadline for the message m. Then value of cumulative distribution at t_d^m gives the probability of meeting the deadline.

$$P\left(t \leq t_d^m\right) = R^m\left(t_d^m\right) \tag{6.14}$$

6.3.1.3 Example

To illustrate the method, benchmark message set of Society of Automotive Engineers (SAE) [13] is considered. The message set have messages exchanged between seven different subsystems in a prototype electric car. The list of messages along with other details are shown in Table 6.2 Message ID 17 is highest priority while message ID 1 is lowest priority. MATLAB code for the example is given in Appendix A.

Using the proposed method, parameters are calculated for each message considering worst-case transmission time. CAN operating speed is 125 kbps (bit-time = 7.745 μs) [13]. For *pdf* ($B_{hp}(t)$ and $t_{rdy}(t)$ estimation number of

Table 6.3 Calculated parameter values

Priority (ID)	bFree	arbSuc	i (No of attempts)
17	0.2039	1.0000	1
16	0.3150	1.0000	1
15	0.2995	0.9986	2
14	0.3150	0.9973	2
13	0.2995	0.9959	3
12	0.3150	0.9946	3
11	0.2902	0.9932	3
10	0.2515	0.9925	3
9	0.2592	0.9918	3
8	0.2592	0.9912	3
7	0.2082	0.9905	3
6	0.2106	0.9904	3
5	0.2082	0.9904	3
4	0.2082	0.9903	3
3	0.2041	0.9902	3
2	0.2039	0.9902	3
1	0.2039	0.9902	3

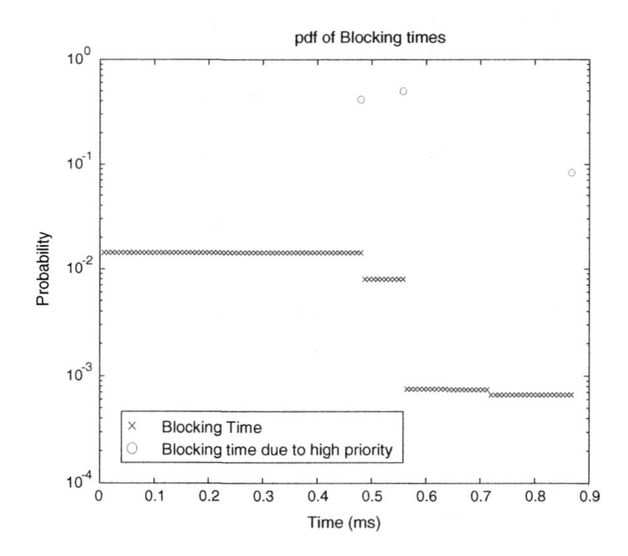

Fig. 6.10 pdf of blocking times

attempts i is truncated such that accumulated probability is ≥ 0.9999. Parameter values $P^m_{\text{free}}, P^m_{\text{high}}$ and number of attempts from *mesReady* are given in Table 6.3 The probability density function of blocking time and blocking time by high priority message, random variables for message ID 9 is shown in Fig. 6.10

Once all the parameter values are available, response-time distribution is estimated.

The response-time distribution of three messages (message ID = 1,9 and 17) is shown in Fig. 6.11. From the response-time distribution, probability of meeting

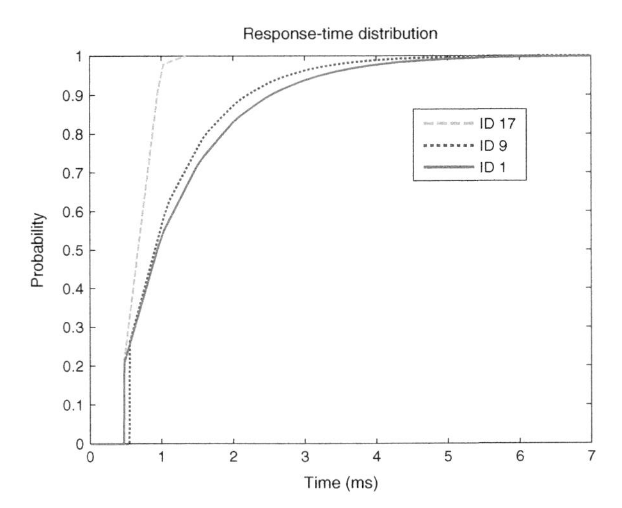

Fig. 6.11 Response-time distribution

Table 6.4 Comparison of response-time results of analysis with literature

Priority (ID)	R_i [13]	$P(R_i)$	R_i^{sim}[13]	$P(R_i\text{sim})$	t_{min}
17	1.416	1.0000	0.680	0.5123	1.324
16	2.016	1.0000	1.240	0.9883	1.402
15	2.536	0.9997	1.720	0.9920	2.238
14	3.136	0.9997	2.280	0.9956	2.742
13	3.656	0.9995	2.760	0.9959	3.369
12	4.256	0.9995	3.320	0.9967	3.919
11	5.016	0.9991	4.184	0.9965	4.957
10	8.376	1.0000	4.664	0.9968	5.553
9	8.976	1.0000	5.224	0.9976	5.925
8	9.576	1.0000	8.424	0.9999	6.374
7	10.096	1.0000	8.904	0.9999	6.831
6	19.096	1.0000	9.616	0.9999	7.094
5	19.616	1.0000	10.096	1.0000	6.940
4	20.136	1.0000	18.952	1.0000	6.978
3	28.976	1.0000	18.952	1.0000	7.172
2	29.496	1.0000	19.432	1.0000	7.025
1	29.520	1.0000	19.912	1.0000	7.032

Second and fourth column give worst-case response-time from literature. Third and fifth column gives probability of message delivery by corresponding time given in second and forth column, respectively. Last column gives response-time from present analysis assuming probability of message delivery to be 0.999

two worst-case times R_i and R_i^{sim} [13] is evaluated. R_i is the worst-case response-time from analysis while R_i^{sim} is from simulation. Columns 2 and 4 in Table 6.4 give these values, corresponding probabilities from response-time distribution analysis is given in columns 3 and 5. Let for all messages, probability of meeting a response-time is fixed to 0.999, then corresponding time value from response-time distribution is given in column 6.

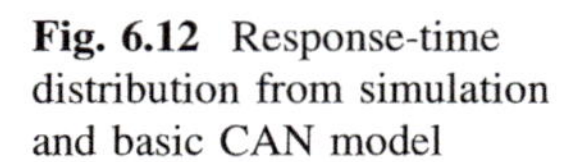

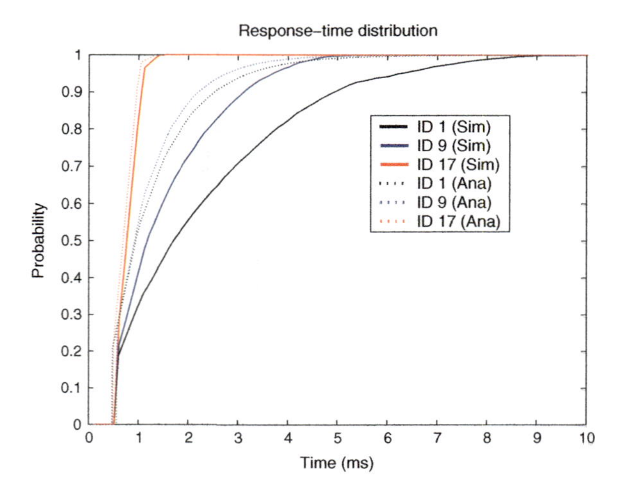

Fig. 6.12 Response-time distribution from simulation and basic CAN model

In Fig. 6.11 the offset at time axis is due to blocking when the message is queued. It is same for all messages irrespective of message priority, because CAN message transmissions are non-preemptive. Slope of response-time curves are different. Slopes are dependent upon the message priority, higher the message priority higher is the slope.

Response-times from worst-case analysis are giving upper bound on response-time, so probability at these times from response-time distribution is expected to be very high or even 1. Values in column 3 of Table 6.4 confirms this. Worst-case response-time from simulation is obtained from a limited simulation (2,000,000 ms [13]). Hence there is no consistence probability at these response-times.

Response-time of message with probability 0.999, is comparable for higher priority messages, while it is almost 25% of worst-case for lower priority. This is because worst-case analysis assumes all higher priority message will get queued deterministically, while response-time distribution gives probabilistic treatment to this.

6.3.1.4 Simulation Model

A simulation model is made using event-triggered approach [14]. Starting time of each message is chosen based on a uniformly distributed random number. During simulation no drift or relative variation among the nodes clock is assumed. Each run is up to 10,000 s. This kind of 10,000 run are simulated to get response time distribution of each message. MALTAB code of the simulation program is given in Appendix A.

Using the simulation model, response-time distribution of messages of SAE CAN example are estimated. The response-time distribution from simulation is plotted along with response-time distribution of corresponding message from basic model in Fig. 6.12.

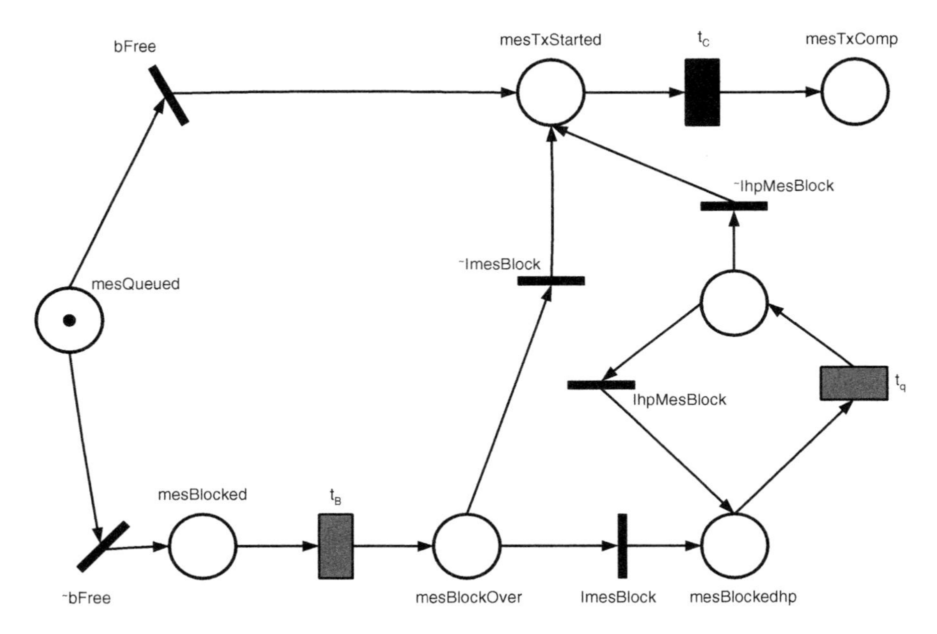

Fig. 6.13 Improved response-time model for CAN

6.3.1.5 Improved Model

Basic CAN model of Fig. 6.9 is analyzed again, in view of the simulation results. It has been found that probability of collision in collision window is almost negligible for all the messages. So, considering this is of not much importance. In basic CAN model, time taken to evaluate probabilities *mesBlock* and *hpmesBlock* is mean of blocking times t_B and t_q, respectively. So, this time is changed from mean to maximum blocking time in the improved model. The improved CAN model is shown in Fig. 6.13. Computation of parameters *ImesBlock* and *IhpmesBlock* is as below:

Probability of no new higher priority message arrival in $t_q(\sim hpMesBlock)$: in the improved mode maximum of BlockTime is used.

$$P_{T_B}^m = \prod_{\substack{i \in M \\ i \in hp(m)}} P_{T_B}^i \tag{6.15}$$

where

$$\begin{aligned} P_{T_B}^i &= \text{Prob}[\text{non occurrence of ith message in time } \mathit{BlockingTime}] \\ &= 1 - \left(\frac{1}{T_i} \times \max[t_B] \right) \end{aligned}$$

Probability of no new higher priority message arrival in $t_q(\sim hpMesBlock)$: This is similar to previous.

$$P^m_{T_{Bhp}} = \prod_{\substack{i \in M \\ i \in hp(m)}} P^i_{T_{Bhp}} \tag{6.16}$$

where

$$\begin{aligned} P^i_{T_{Bhp}} &= \text{Prob}[\text{non occurrence of ith message in time } BlockTimebyNew] \\ &= 1 - \left(\frac{1}{T_i} \times \max[t_{Bhp}]\right) \end{aligned}$$

Queueing time: time to reach place *mesTxStarted* from *mesBlockedhp* in *i*th step is modeled as a single r.v. with *pdf* $B^m_{hp}(i,t)$.

$$B^m_{hp}(i,t) = \left[\left(1 - P^m_{T_{Bhp}}\right)^{i-1} P^m_{T_{Bhp}}\right] \eta^i(t)$$

where

$$\begin{aligned} \eta^i(t) &= \eta^{i-1}(t) \otimes p^m_{bhp}(t) \\ \eta(t) &= p^m_{bhp}(t) \\ B^m_{hp}(t) &= \sum_i B_{hp^m}(i,t) \end{aligned}$$

In the improved model, token from place *mesQueued* can reach place *mesTxStarted* by following either of these following paths, (1) directly by firing of transition *bFree*, (2) firing of transitions $bFree, t_B$, *ImesBlock* and, (3) firing of transitions $\sim bFree, t_B, ImesBlock$, loop of transitions $t_q\, IhpMesBlock$ and escaping transition $\sim IhpMesBlock$. Total queueing time for improved model is given as:

$$\begin{aligned} q^m(t) = {} & P^m_{\text{free}}\delta(t) + \left(1 - P^m_{\text{free}}\right) P^m_{T_B} p^m_b(t) \\ & + \left(1 - P^m_{\text{free}}\right)\left(1 - P^m_{T_B}\right)[p^m_b(t) \otimes B^m_{hp}(t)] \end{aligned} \tag{6.17}$$

Response-time distribution for CAN message set of Table 6.2 is evaluated using improved CAN model. MATLAB code of the improved CAN model is given in Appendix A.

The response-time distribution for three message IDs 1, 9 and 17 from improved CAN model is plotted along with simulation results in Fig. 6.14.

Here both response-time distributions seems to be in quite good agreement at higher probability values.

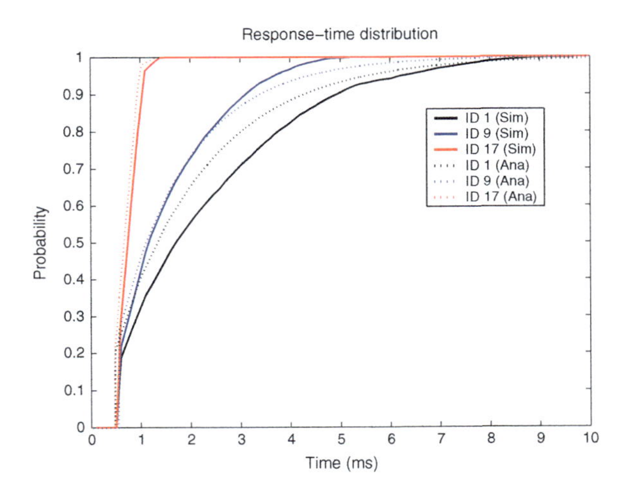

Fig. 6.14 Response-time distribution from simulation and improved CAN model

6.3.1.6 Effect of Bit-Stuffing

Total number of stuffed bits in a message depends on bit pattern of message. A probabilistic treatment requires probability of each bit pattern to get distribution of bit-stuffing for each message as given in [13, 15].

Once the distribution of number of stuff bits is available, distribution of transmission time can be obtained [13]

$$C_m(t) = C_m + (\varphi(n))\tau_{\text{bit}} \tag{6.18}$$

where

C_m time taken to transmit date without any stuff bits
$\varphi(n)$ prob. that stuff bits are n
τ_{bit} time taken to transmit a bit on the bus

Response-time of a message considering bit-stuffing can be determined using above method, by replacing fixed C_i by distribution $C_i(t)$ in (6.6–6.8).

CAN message might get corrupted due to interference of EMI [16, 17]. The transmitting node have to retransmit the message, thus, corruption of message have effect of increasing message's network delay. The interference affects *probability of finding free* and both the *blocking times* of the CAN model in addition to retransmission.

6.3.2 MIL-STD-1553B

6.3.2.1 Worst-Case Delay Analysis

As per [18], the delay for cyclic service network can be simply modeled as a periodic function such that $\tau_k^{\text{SC}} = \tau_{k+1}^{\text{SC}}$ and $\tau_k^{\text{CA}} = \tau_{k+1}^{\text{CA}}$ where τ_k^{SC} and τ_k^{CA} are the

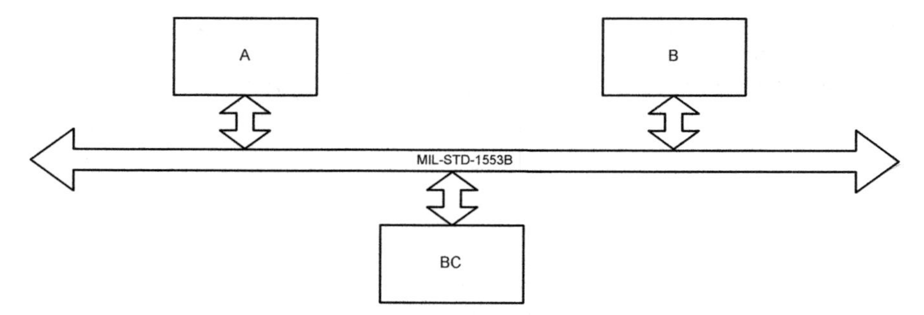

Fig. 6.15 Two nodes of a MIL-STD-1553B network

sensor-to-controller delay and the controller-to-actuator delay at sampling time period k. The model works perfectly in ideal cases, i.e. when local clock are very stable. In practice, systems may experience small variations on periodic delays due to several reasons. For example, drift in clock generators of nodes & controllers, failure & restoration of node may result in delay variation.

6.3.2.2 Response Time Model

Consider a network with two nodes as shown in Fig. 6.15. BC of the network controls transfer of data on network. The network delay for data transfer from node A to B is defined as

$$\tau_{\mathrm{AB}} = t^{A}_{\mathrm{suc}} - t^{A}_{Q} \tag{6.19}$$

In this equation τ_{AB} is the network delay experienced by message at node A for transfer to node B. t^{A}_{suc} is the time of successful transfer of data from node A, t^{A}_{Q} is the time of queuing of data by node A for transfer to B.

Node A is allowed to transmit its data to B periodically under the command of BC. As node A and BC are not synchronized, waiting time (queuing time to the time of actual start of transmission) will have uniform distribution. This uniform distribution has range $\left(0, \tau^{\mathrm{AB}}_{\mathrm{mil}}\right)$. $\tau^{\mathrm{AB}}_{\mathrm{mil}}$ is time period or cycle time of data transfer from A to B. Once node A gets turn for message transfer to B, it starts putting the frame. The transmission delay has two components, frame size and prorogation delay. Frame size is proportional to number of bytes to be transferred, while propagation delay is because media length connecting nodes A and B. Now in terms of waiting time, frame time and propagation time, network delay can be written as:

$$\tau_{\mathrm{AB}} = \tau^{A}_{\mathrm{wait}} + \tau^{A}_{fr} + \tau^{A}_{\mathrm{prop}} \tag{6.20}$$

For a given data and pair of nodes framing time and propagation time are constant. The sum of these two is referred as transmission time. With the assumption that data is not corrupted during framing or propagation (i.e no retransmission), transmission time is constant. So, network delay is sum of a random (waiting time) and fixed (transmission time) quantity.

Let transmission time of the message transmitted from node A to B is denoted as τ_{tx}^{AB}. Then *pdf* of transmission delay is given as:

$$d^{\mathrm{AB}}(t) = \frac{1}{\tau_{\mathrm{mil}}^{\mathrm{AB}}} \int_{0}^{t} \left[U(\tau) - U\left(\tau - \tau_{\mathrm{mil}}^{\mathrm{AB}}\right)\right]\delta\left(t - \tau + \tau_{tx}^{\mathrm{AB}}\right)d\tau \qquad (6.21)$$

From (6.21), it is clear under the assumption of no retransmission, network delay has time shifted uniform distribution. This time shift is by fixed transmission time.

Effect of EMI is slightly different from that in CAN. Transmitter might retransmit the frame again or wait for next cycle, based on the implementation. So, interference might result in network delay or message loss as well.

6.3.3 Ethernet

We will start with a myths about Ethernet performance. The main is that the Ethernet system saturates at 37% utilization. The 37% number was first reported by Bob Metcalfe and David Boggs in their 1976 paper titled "Ethernet: Distributed Packet Switching for Local Computer Networks". The paper considered a simple model of Ethernet. The model used the smallest frame size and assumed a constantly transmitting set of 256 stations. With this model, the system reached saturation at $1/e$ (36.8%) channel utilization. The authors had warned that this was simplified model of a constantly overloaded channel, and did not bear any relationship to the normal functioning networks. But the myth is persistent that Ethernet saturates at 37% load.

Ethernet is a 1-persistent CSMA/CD protocol. 1-persistent means, a node with ready data senses the bus for carrier, and try to acquire the bus as soon as it becomes free. This protocol has serious problem, following a collision detection. Nodes involved in collision, keep on colliding with this 1-persistent behavior. To resolve this, Ethernet has BEB (binary exponential backoff) algorithm. Incorporation of BEB introduces random delays and makes the response time indeterministic.

In this section, we proposes a method to estimate response time distribution for a given system. The method is based on stochastic processes and operations. The method requires definition of model parameters.

6.3.3.1 Response-Time Model

A DSPN model for response-time modeling of a message in basic Ethernet (10/100 Mbps) is given in Fig. 6.16.

A token in place *mesReady* in Fig. 6.16 depicts a massage ready for transmission at a node. Place *BusFree* stands for the condition when node senses that bus (i.e. physical medium) is free. *BusAcq* stands for the condition when message transmission is in progress in excess to minimum packet size. Place named *Collision* depict the collision on the bus. Places *mesBackedOff* and

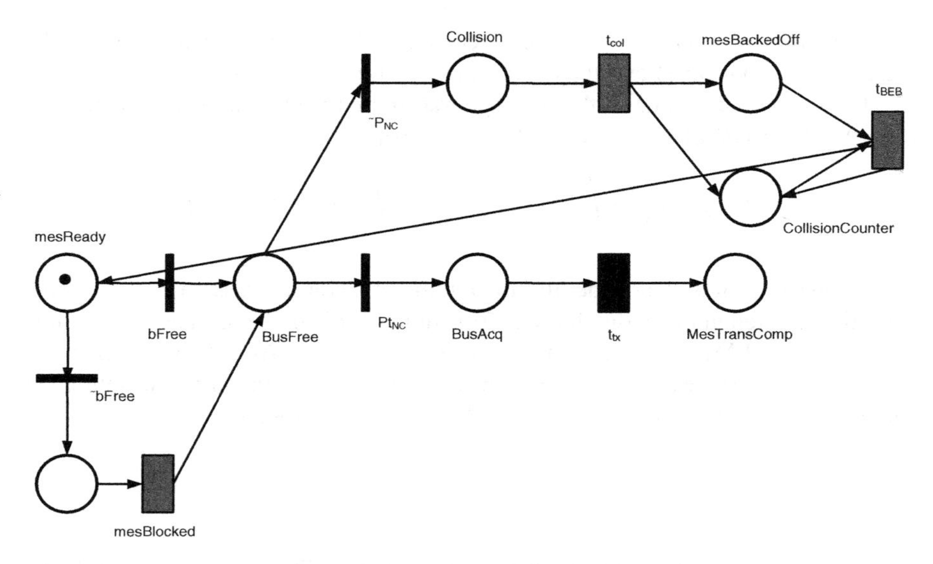

Fig. 6.16 Response-time model for messages on basic Ethernet

CollisionCounter depicts the back off condition and retransmission counter, respectively. There are four immediate transitions – $bfree, \tilde{bfree}, P_{tNC}, \tilde{P}_{tNC}$ – in the model. The two are actually complementary of each other. Transitions $bFree, P_{tNC}$ depicts the probability of finding the bus free and probability of no collision, respectively. Timed transition t_{tx} represents the fixed time of transmission of the message. General timed transitions $mesBlocked, t_{\mathrm{col}}, t_{\mathrm{BEB}}$ represents blocking time when bus is not free, collision time and variable back off time, respectively.

The model counts the number of collisions (or retries) before actual transmission. t_{BEB} need to be varied as the count changes. It does not model stoppage of transmission after 16 attempts.

The estimation of these probabilities and general time distribution is the most difficult part. It varies with number of message ready for transmission and other network parameters.

6.4 System Response-Time Models

6.4.1 Sample to Actuation Delay and Response-Time

In networked system, two parameter of importance are sample to actuation delay and response-time. First is the time difference of actuating action to the corresponding sampling time. While the second is time taken by networked system to react (or respond) to an action of physical process.

Assuming that all nodes are time-triggered, i.e. they sample inputs from physical process or network periodically. Nodes generate output to physical process or network after computation time. For a node i, sampling period is denoted by τ^i_{samp} and delay after sampling is computation delay is denoted as τ^i_{comp}.

In case of networked system, sampling of process input is carried out by sensor node and actuation for corrective action is done by actuator node. So, sample to actuation delay is given by:

$$\tau_{a-s} = \tau^S_{\text{comp}} + \tau_{\text{SC}} + \tau^C_{\text{samp}} + \tau^C_{\text{comp}} + \tau_{\text{CA}} + \tau^A_{\text{samp}} + \tau^A_{\text{comp}} \tag{6.22}$$

Computation delay is negligible for some nodes (sensor, actuator). For controller nodes it is finite. For analysis purpose it can be assumed constant. Sampling time due to phase difference is modeled as uniform distribution.

$$\tau^i_{\text{samp}} \models \text{Unif}\left(0, t^i_{\text{samp}}\right) \tag{6.23}$$

So, *pdf* of sample to actuation delay is convolution of all the variable of above equation:

$$\begin{aligned} d^{a-s}(t) &= d^{\text{SC}}(t) \otimes \text{unif}\left(0, t^C_{\text{samp}}\right) \\ &\otimes \delta\left(t - t^C_{\text{comp}}\right) \otimes d^{\text{CA}}(t) \\ &\otimes \text{unif}\left(0, t^A_{\text{samp}}\right) \end{aligned} \tag{6.24}$$

Since sensor node samples the physical inputs periodically, response-time density is given as:

$$r(t) = \text{unif}\left(0, t^S_{\text{samp}}\right) \otimes d^{a-s}(t) \tag{6.25}$$

6.4.1.1 Example: Numerical Example—Deterministic Case

Take a distributed system as shown in Fig. 6.17. System is configured as TMR (Triple Modular Redundant). It consists of three input nodes ($IPN_{A/B/C}$), three output nodes ($OPN_{A/B/C}$), and one communication link or channel, C_1. The behavior of the node and communication channel is as follows:

IPN's run a cyclic program of acquiring primary inputs, processing them as per the defined algorithm (or logic) and put output of processing in the buffer of its communication interface for transmission to respective OPN (i.e. IPN_A to OPN_A). The node may have any of the scheduling algorithms such as, round-robin (RR), priority with preemption, and priority without preemption etc. and load on the system will decide the periodicity in acquiring primary inputs and also the time required to perform processing. Time elapsed from the instant of a change in input to its acquisition is a random variable denoted as τ^I_{acq} with domain (0, cycle time of IPN). Similarly, time required for processing and depositing output to the buffer of communication interface is also a random variable, τ^I_{pro} with domain $\Re$(set of +ve real numbers).

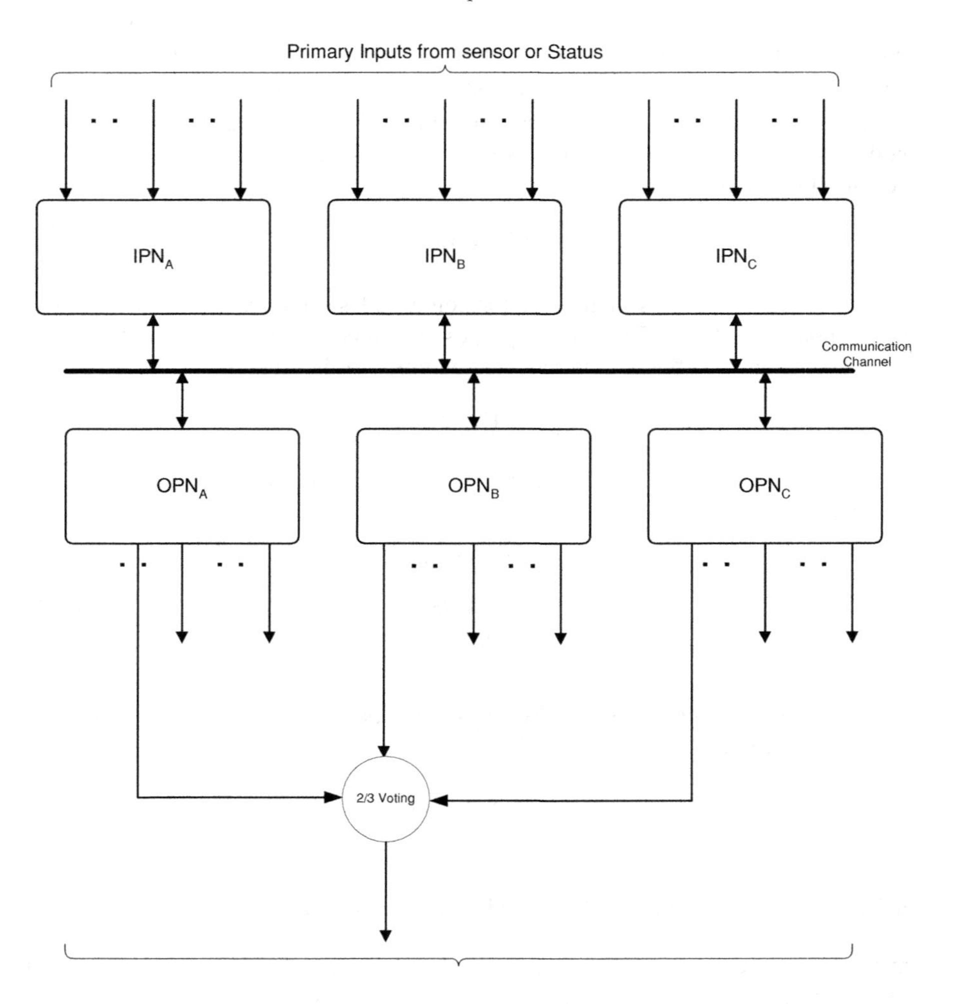

Fig. 6.17 Distributed system with triple redundant nodes

OPN's also run a cyclic program of acquiring outputs from their respective IPN's, and generates commands for the actuators/manipulators. Similar to IPNs, time elapsed from the instant of reception of commands at communication interface to its acquisition is a random variable τ^{O}_{acq} with domain (0, cycle time of OPN) and time required for generation of command for actuators, τ^{O}_{gen} with domain $\Re$.

Let us take a deterministic case. Here all the random variable are having fixed values. The density function of fixed random variables is denoted by *Dirac delta* function, $\delta(t)$. Response-time density functions of random variable of channel A are given in Table 6.5.

Table 6.5 Values of random variables

Notation	Value	Density function
$\dot{F}^1(t)$	3	$\delta(t-3)$
$\dot{F}^2(t)$	2	$\delta(t-2)$
$\dot{F}^3(t)$	5	$\delta(t-5)$
$\dot{F}^4(t)$	3	$\delta(t-3)$
$\dot{F}^5(t)$	1	$\delta(t-1)$

Using (3–3) channel A response-time density is given as:

$$\dot{F}^A(\tau) = \dot{F}^1 \otimes \dot{F}^2 \otimes \dot{F}^3 \otimes \dot{F}^4 \otimes \dot{F}^5(\tau)$$

putting values of individual response-time densities in above equation, we get:

$$F^A(t) = \int_0^t \int_0^{\tau_5} \int_0^{\tau_4} \int_0^{\tau_3} \int_0^{\tau_2} \delta(\tau_1 - 3)\delta(\tau_2 - \tau_1 - 2)\delta(\tau_3 - \tau_2 - 5)\delta(\tau_4 - \tau_3 - 3) \times \delta(\tau_5 - \tau_4 - 1)d\tau_1 d\tau_2 d\tau_3 d\tau_4 d\tau_5$$

Solving this for response-time distribution, gives the following:

$$F^A(t) = \int_0^t \dot{F}^A(\tau)d\tau = \cup(t-14)$$

In similar manner, response-time distribution of other two channels, B and C can be obtained. Let's take following:

$$F^B(t) = \int_0^t \dot{F}^B(\tau)d\tau = \cup(t-15)$$

and

$$F^C(t) = \int_0^t \dot{F}^C(\tau)d\tau = \cup(t-16)$$

Using (3–8), response-time distribution of system can be obtained. Mathematically it is given as:

$$F^{Sys}(t) = \begin{Bmatrix} \cup(t-14)\cup(t-15) \\ +\cup(t-15)\cup(t-16) \\ +\cup(t-16)\cup(t-14) \end{Bmatrix} - \begin{Bmatrix} \cup(t-14)\cup(t-15)\cup(t-15)\cup(t-16) \\ +\cup(t-14)\cup(t-15)\cup(t-16)\cup(t-14) \\ +\cup(t-15)\cup(t-16)\cup(t-16)\cup(t-14) \end{Bmatrix}$$

$$+\begin{Bmatrix} \cup(t-14)\cup(t-15)\cup(t-15)\cup(t-16)\cup(t-16)\cup(t-14) \\ \cup(t-14)\cup(t-15)\cup(t-15)\cup(t-16)\cup(t-16)\cup(t-14) \end{Bmatrix}$$

Table 6.6 Notations and distributions of random variables

Notation	Mean value	Distribution	Density function
$\dot{F}^1(t)$	3	Uniform	$\frac{1}{6}(\mathrm{U}(t) - \mathrm{U}(t-6))$
$\dot{F}^2(t)$	2	Exponential	$\frac{1}{2}e^{\frac{-t}{2}}$
$\dot{F}^3(t)$	5	Exponential	$\frac{1}{5}e^{\frac{-t}{5}}$
$\dot{F}^4(t)$	3	Uniform	$\frac{1}{6}(\mathrm{U}(t) - \mathrm{U}(t-6))$
$\dot{F}^5(t)$	1	Exponential	e^{-t}

The result obtained needs no explanation; it is self-evident.

$$F^{Sys}(t) = \mathrm{U}(t-15)$$

6.4.1.2 Example: Numerical Example—Non-Deterministic Case

In this case random variables for time to acquire inputs are assumed to be uniformly distributed over its cycle time. This assumption is based on the fact that instant of change of input and instant of acquisitions are statistically independent. Random variable for time to process (or transmit/generate) at IPN (communication channel/OPN) is assumed to be exponentially distributed. This assumption is based on the fact that these systems (IPN/OPN/communication channel) are having complex interactions within them, making then non-deterministic and memoryless. Memoryless means, remaining time to process does not depend on how long it is being processed. Memory less random variable in continuous domain is exponential.

Density functions of the random variables are given in Table 6.6.

Using (3–3) and (3–4), response-time densities and distributions can be obtained. The response-time distribution for a channel, e.g. channel A is given as:

$$\begin{aligned} F^A(t) = & -2/9 \times t + 1/72 \times U(t-6) \times \exp(-t+6) - 625/432 \times U(t-12) \\ & \times \exp(-1/5 \times t + 12/5) - 8/27 \times U(t-6) \times \exp(-1/2 \times t + 3) \\ & - 5/9 \times U(t-12) \times t + 625/216 \times U(t-6) \times \exp(-1/5 \times t + 6/5) \\ & - 113/18 \times U(t-6) + 4/27 \times U(t-12) \times \exp(-1/2 \times t + 6) \\ & - 1/144 \times U(t-12) \times \exp(12-t) + 4/27 \times \exp(-1/2 \times t) - 1/144 \\ & \times \exp(-t) + 1/72 \times t^2 + 215/36 \times U(t-12) + 7/9 \times U(t-6) \\ & \times t + 1/72 \times U(t-12) \times t^2 - 625/432 \times \exp(-1/5 \times t) - 1/36 \\ & \times U(t-6) \times t^2 + 47/36 \end{aligned}$$

Assuming channel response-time distribution for all the three channels same, system response-time distribution can be calculated using (3–8).

$$F^{\mathrm{Sys}}(t) = 3 * (1/72 \times t^2 - 2/9 \times t - 113/18 \times U(t-6) + 4/27 \times U(t-12) \times \exp(-1/2 \times t + 6) - 625/432 \times \exp(-1/5 \times t) - 8/27 \times U(t-6) \times \exp(-1/2 \times t + 3) - 1/144 \times \exp(-t) - 625/432 \times U(t-12) \times \exp(-1/5 \times t + 12/5) + 215/36 \times U(t-12) - 1/144 \times U(t-12) \times \exp(12-t) + 1/72 \times U(t-6) \times \exp(-t+6) + 625/216 \times U(t-6) \times \exp(-1/5 \times t + 6/5) + 4/27 \times \exp(-1/2 \times t) + 1/72 \times U(t-12) \times t^2 + 7/9 \times U(t-6) \times t - 1/36 \times U(t-6) \times t^2 - 5/9 \times U(t-12) \times t + 47/36)^2 - 3 \times (1/72 \times t^2 - 2/9 \times t - 113/18 \times U(t-6) + 4/27 \times U(t-12) \times \exp(-1/2 \times t + 6) - 625/432 \times \exp(-1/5 \times t) - 8/27 \times U(t-6) \times \exp(-1/2 \times t + 3) - 1/144 \times \exp(-t) - 625/432 \times U(t-12) \times \exp(-1/5 \times t + 12/5) + 215/36 \times U(t-12) - 1/144 \times U(t-12) \times \exp(12-t) + 1/72 \times U(t-6) \times \exp(-t+6) + 625/216 \times U(t-6) \times \exp(-1/5 \times t + 6/5) + 4/27 \times \exp(-1/2 \times t) + 1/72 \times U(t-12) \times t^2 + 7/9 \times U(t-6) \times t - 1/36 \times U(t-6) \times t^2 - 5/9 \times U(t-12) \times t + 47/36)^4 + (1/72 \times t^2 - 2/9 \times t - 113/18 \times U(t-6) + 4/27 \times U(t-12) \times \exp(-1/2 \times t + 6) - 625/432 \times \exp(-1/5 \times t) - 8/27 \times U(t-6) \times \exp(-1/2 \times t + 3) - 1/144 \times \exp(-t) - 625/432 \times U(t-12) \times \exp(-1/5 \times t + 12/5) + 215/36 \times U(t-12) - 1/144 \times U(t-12) \times \exp(12-t) + 1/72 \times U(t-6) \times \exp(-t+6) + 625/216 \times U(t-6) \times \exp(-1/5 \times t + 6/5) + 4/27 \times \exp(-1/2 \times t) + 1/72 \times U(t-12) \times t^2 + 7/9 \times U(t-6) \times t - 1/36 \times U(t-6) \times t^2 - 5/9 \times U(t-12) \times t + 47/36)^6$$

Channel response-time density, distribution and system response-time distributions are plotted in Fig. 6.18.

6.4.2 *Effect of Node Redundancy*

In networked systems, to make the system fault-tolerant, it's node group (sensor, controller and actuator) can have redundancy. This redundancy could of any form and type, active/passive, hot, cold or warm [19], etc. Here MooN (M-out-of-N) redundancy is considered. A generic diagram is shown in Fig 6.19.

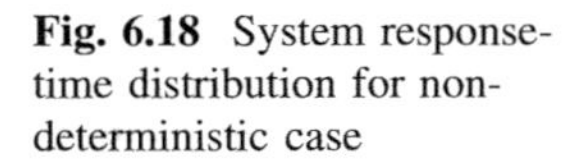

Fig. 6.18 System response-time distribution for non-deterministic case

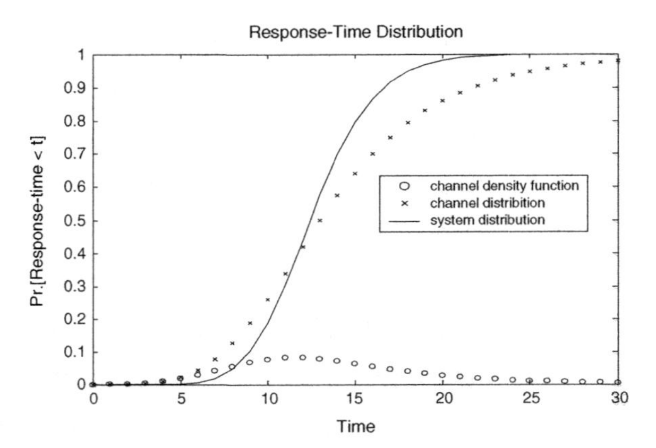

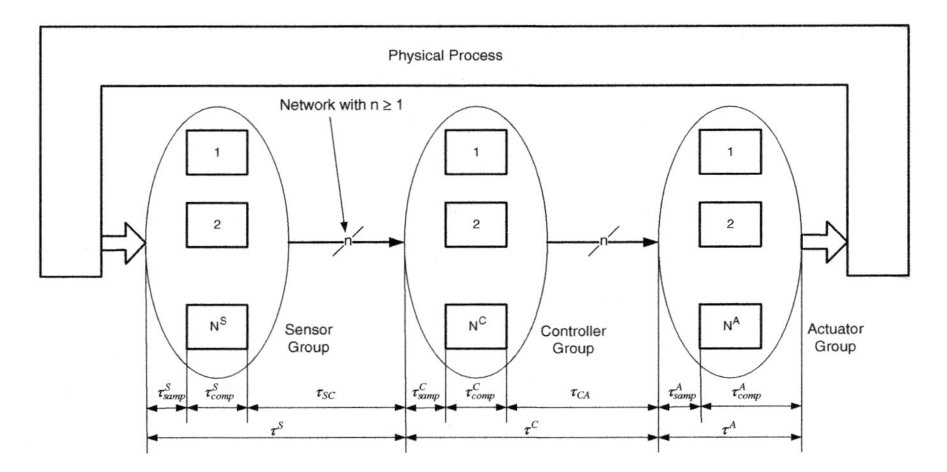

Fig. 6.19 NCS response-time with redundancies

All nodes in a group are independent. Each sensor node of sensor group sample the physical parameter independently. Parameters from network are common to all the nodes of a group. For parameters from network, nodes might wait for copies of same message from multiple nodes of the sending group. For example, in case sensor group is having triple redundancy, controller nodes might wait for a data from two different nodes of sensor group for processing for majority voting system. As all nodes are time-triggered, a controller node performs $2oo3$ of recent messages from sensor group.

Let redundancy of each group is denoted as $M^i ooN^i$ and $i \in \{S, C, A\}$. Receiving nodes of a data wait for data from sending group, i from at least M_i nodes. To model response-time in this scenario, we define group delay time. Group delay time of a group j, is defined as time difference of successful transmission of output data by M_j nodes from occurrence of an event at physical process or receipt of data at nodes of group j. Where i, is the sending group.

$$\tau^i = \left(m^i\right)\text{th}\, median \left\{ \begin{array}{l} \left(\tau^i_{\text{samp}} + \tau^i_{\text{comp}} + \tau_{ij}\right)_1, \ldots, \\ \left(\tau^i_{\text{samp}} + \tau^i_{\text{comp}} + \tau_{ij}\right)_{N^i} \end{array} \right\} \tag{6.26}$$

This is again an order statistics problem. As sample, compute and network delay of each node independent and identically distributed $(i \cdot i \cdot d \cdot)$, $d^i_1(t)$ is used to denote *pdf* of this sum. Corresponding distribution function is denoted as $D^i_1(t)$. So, group delay is given as:

$$D^i(t) = \sum_{n=M^i}^{N^i} \left[D^i_1(t)\right]^n \left[1 - D^i_1(t)\right]^{N^i - n} \tag{6.27}$$

With this response-time model of the system in Fig. 6.19 is given as:

$$r(t) = d^S(t) \otimes d^C(t) \otimes d^A(t) \tag{6.28}$$

6.4.3 Dependence of Response-Time Between Consecutive Cycles

In the above analysis, we considered that nodes are asynchronous. The analysis is valid only for initial or first response, as nodes and network controllers are carrying out their periodically. Local clocks of nodes as well as network controllers (if any) are very stable. So, if system response-time is, x at first cycle after startup, chances are very high that it will remain x in consecutive cycle(s). Response-times of consecutive cycles are correlated. From common logic it can be inferred, any variation in consecutive response-times will be due to drift in clocks, failure of node/controller and restoration of node/controller.

A networked real-time system, able to meets its deadline at startup may fail after operating for time t, due to drift in clocks, failure and restoration of nodes/network controllers.

6.4.3.1 CAN

Network delay on CAN has variation mainly due to traffic and priority among messages. Message with lower priority has higher variation than high priority messages. Network delay of a message in one cycle is not related with the delay in the previous cycle, as the interacting traffic is independent. So, network delay in each cycle is independent and follow the *pdf* given by (6.12).

6.4.3.2 MIL-STD-1553B

As discussed earlier, network delay in case of MIL-STD-1553B has variation mainly because of mismatch in clocks of remote terminals and bus controllers. Most of the cases, this mismatch is very small. So, network delay in consecutive cycles will have small variations, means network delay is dependent on previous cycle. Let network delay of data from node A to B is measured in cycle k, then network delay in cycle $k+1$ is going to be around the measured value with a high probability. Although network response-time at any random cycle will follow the *pdf* given by (6.21). But network delay in consecutive cycle will be dependent on current cycle network delay.

Let p_{pre} is the probability that network delay in any cycle is same as in the previous cycle, then network delay in successive cycles is given as:

$$d^{\text{AB}}(t,i) = \delta(t-x) \tag{6.29}$$

where x is the network delay in cycle i

$$d^{\text{AB}}(t,i+1) = p_{\text{pre}}\delta(t-x) + \left(1-p_{\text{pre}}\right)d^{\text{AB}}(t) \tag{6.30}$$

and

$$\begin{aligned} d^{\text{AB}}(t,i+2) &= p_{\text{pre}}d^{\text{AB}}(t,i+1) + \left(1-p_{\text{pre}}\right)d^{\text{AB}}(t) \\ &= p_{\text{pre}}^{2}\delta(t-x) + \left(1-p_{\text{pre}}\right)\left(1+p_{\text{pre}}\right)d^{\text{AB}}(t) \end{aligned}$$

similarly

$$\begin{aligned} d^{\text{AB}}(t,i+n) &= p_{\text{pre}}d^{\text{AB}}(t,n-1) + \left(1-p_{\text{pre}}\right)d^{\text{AB}}(t) \\ &= p_{\text{pre}}^{n}\delta(t-x) + \left(1-p_{\text{pre}}^{n}\right)d^{\text{AB}}(t) \end{aligned} \tag{6.31}$$

If $p_{\text{pre}} = 1$, i.e. there is no mismatch in clocks, then nth cycle will also have the same network delay as ith cycle. When $p_{\text{pre}} < 1$, then network delay in nth cycle will be given be $d^{\text{AB}}(t)$.

Similarly, if the receiving node has received the message in present cycle before a specified time, t_{α}, then the distribution of delay time in next cycle is given be the above equations. Let conditional density is denoted as $d^{\text{AB}}(t|t_{\alpha})$. Dirac delta function in above equations is replaced by conditional density function.

6.4.4 Failure/Repair Within the System

Node and network channel may fail, repaired and restored back to operation. When a node or network channel is not available because of failure, it might affect the system response-time distribution and probability of meeting deadline(s).

Assuming link, i.e. interconnects, failure probability to be negligible, failure location-wise system elements can be categorized as failures in nodes and failures in network. Further, failures in node can be divided as (1) failure in processing unit (PU), (2) network interface unit (NIU) and, (3) complete failure, e.g. power supply, etc. Failure in node's PU might affect data, but failure in NIU and complete failure affect response-time distribution. Network might fail if it has active components such as bus controller (BC)—in case of MIL-STD-1553B—, hubs or switches.

For a NRT system with redundancy at node and network level, the network delay distribution and response-time distribution, for each possible state of failure state-space is evaluated. Number of states will increase rapidly with system size. So to limit the evaluation of distributions for all possible state, distribution in hardware-wise healthy states is considered.

The system will fail, if at any time number of failed nodes in a group exceeds tolerable number of failures. Response-time distribution for a given set of nodes and network channel operating can be evaluated using the already explained steps. Using the steps, response-time distribution for all healthy states can be derived.

Response-time distribution in each system state (i, j, k, n), where i, j, k, n are number of *UP* sensor nodes, controller nodes, actuator nodes and network channels, respectively.

6.5 Timeliness Hazard Rate

Systems usually have robust control algorithm to tolerate message delays and drops. Message delay and drop has effect of delaying control action by networked real-time systems. Delay beyond a specified time in taking control action is considered timeliness failure.

Let system failure criteria is n consecutive deadline violation (or timeliness failures). When $n = 1$, number of cycles at which timeliness failure will occur follows geometric distribution [12].

$$P(Z = i) = p^{i-1}q \tag{6.32}$$

where Z: random variable; q: probability of occurrence of timeliness failure; $p : 1 - q$.

Geometric distribution is a memoryless distribution in discrete time and is counterpart of exponential distribution in continuous time [12]. At gross level (larger time scale), it can be easily converted to exponential distribution. In exponential distribution characterizing parameter is hazard rate, which in this case is referred to as "*timeliness hazard rate*".

$$\lambda^T = \frac{1}{t}\ln\left(\frac{1}{P\left(Z > \left\lceil \frac{t}{t_C} \right\rceil\right)}\right) \tag{6.33}$$

where λ: Timeliness hazard rate; t: Operating time; t: Cycle time.

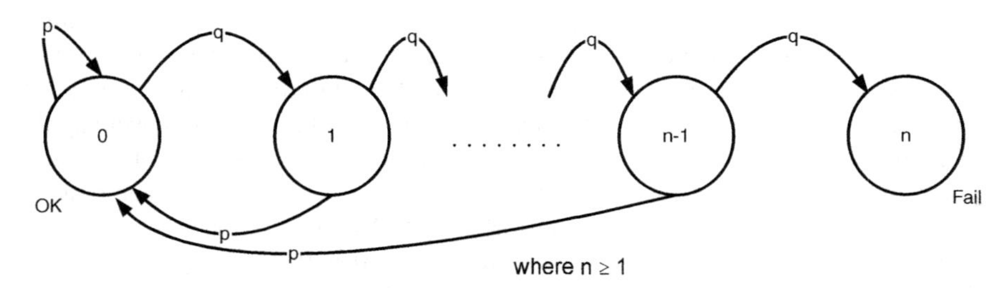

Fig. 6.20 DTMC of system failure due to timeliness

When $n > 1$, number of cycles for timeliness failure will not follow geometric distribution. This process (number of cycles for timeliness failure) is a memory process and directly cannot be modeled as Markov. Using the technique of additional states [20], Markov model can be used to model this process. A DTMC for this process is shown in Fig. 6.20. It is clear a general equation for n cannot be given. This gives rise to computational problem for higher values of n. Also, it will be better if this process can be mapped to a continuous time process, as it will enable modeling of timeliness failure along with hardware failures in system dependability modeling.

One algorithm to evaluate timeliness hazard is by using 6.33. Alternate method to estimate hazard rate is by its definition [12]. Hazard rate at any given time t, is conditional instantaneous probability of failure at time, given that it has survived up to time t [12].

$$\lambda^T(t) = \frac{f(t)}{1-F(t)} = \frac{f(t)}{R(t)} \tag{6.34}$$

where: $F(t)$: Failure distribution (CDF); $f(t)$: Failure density (pdf); $R(t)$: Reliability.

Equation (6.34) requires $F(t)$ as differentiable function. While DTMC will give $F(t)$ values at discrete points only. Using techniques of discrete mathematics, these discrete values are used to evaluate timeliness hazard rate. MATLAB code for deriving timeliness hazard rate is given in Appendix A.

6.5.1 Example 1

Let probability of meeting the specified deadline is $p = 0.99998$ per cycle. And cycle time is 50 ms. Timeliness hazard rate, λ^T for three timing failure criteria, $n = 1, 2$ and 3, as evaluated using (6.33) is given in Table 6.7

Timeliness failure probabilities are estimated using DTMC of Fig. 6.20 and using exponential distribution with hazard rate of Table 6.7

Plot of timeliness hazard rate with time does not show any trend, as shown in Fig. 6.21. All three hazard rates are constant with time, so exponential distribution

Table 6.7 Probability and hazard rates with exponential distribution

Probability	λ_1^T (per cycle)	λ_2^T (per cycle)	λ_3^T (per cycle)
0.99998	2.00E-05	4.00E-10	8.00E-15

For the given probability of meeting deadline in one cycle, hazard rate for three failure conditions, $n = 1$, 2, and 3 is given in second to fourth columns

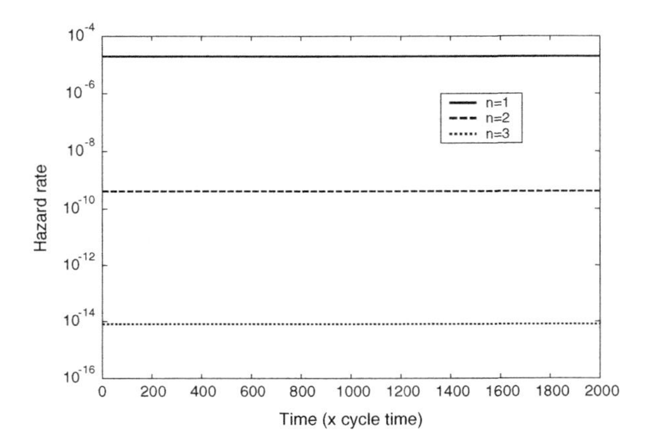

Fig. 6.21 Hazard rates with time

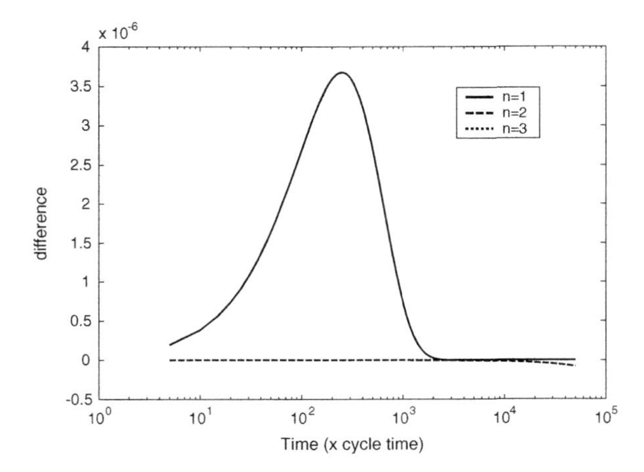

Fig. 6.22 Difference in estimated probabilities with time. The difference is taken as (failure probability from DTMC—failure probability from exponential failure distribution)

can be used to model failure distribution [12, 20]. Difference in estimated probabilities is plotted in Fig. 6.22.

The process with more than one consecutive failure although have memory at micro-scale i.e. at cycle level, but at a larger time scale it may not. If there is no timeliness failure up to time t, then chances of timeliness failure in time $t + \delta t$ in independent of t. Means, system state at t w.r.t. timeliness failures is as good as new.

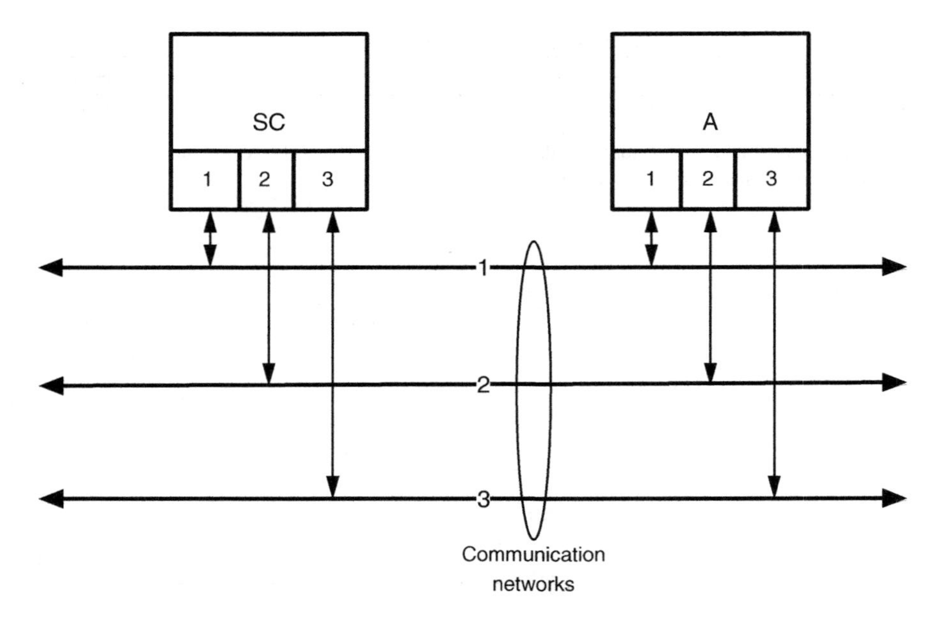

Fig. 6.23 Schematic of example system

6.5.2 Example 2

Consider a two node system with three networks interconnecting them. One node acts as sensor & controller (SC) node and other node acts as actuator (A) node. Interconnecting network system is taken to be CAN in one case and MIL-STD-1553B in the other. The schematic of the system is shown in Fig. 6.23.

System components may fail and restored back by means of repair. In degraded states (system has some faulty components, but still system is functional) response time change. In all working states system's response-time is computed using method discussed earlier in this chapter.

In this example system, message set of previous example is taken. Messages with 2 bytes and 10 ms cycle time are considered for both the cases. On network other messages are taken. System has total of 17 messages as per Table 6.2. Messages for present case I, have message ID 8, 9 and 12 (with message ID 12 having 10 ms as period). For case I following healthy and degraded states are considered:

1. All nodes are healthy and at least one network is healthy
2. One SC node is Down and at least one network is healthy
3. One A node is Down and at least one network is healthy
4. One SC and one A nodes are Down and at least one network is healthy

In case of CAN bus, redundancy at CAN level does not affect the system response-time, while in case of MIL-STD-1553B, failure of network channels do

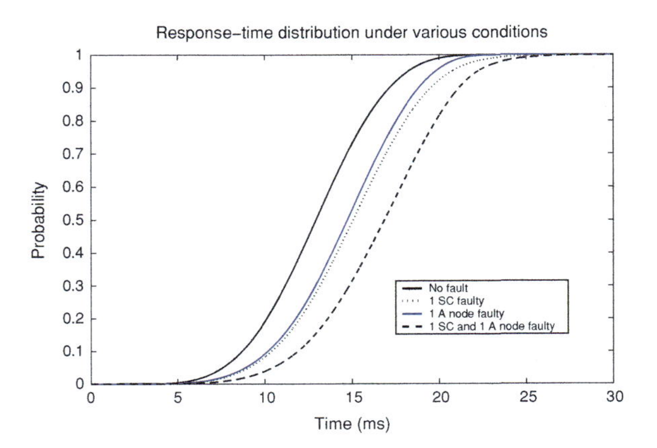

Fig. 6.24 Response-time distribution under various operating states of case I

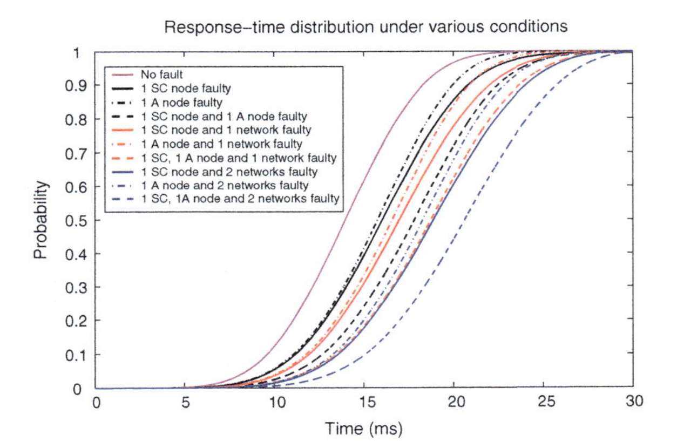

Fig. 6.25 Response-time distribution under various operating states of case II

affect the system response-time. For case II following healthy and degraded states are considered:

1. All nodes and network channels are healthy
2. One SC node is Down and all network channels are healthy
3. One A node is Down and all network channels are healthy
4. One SC and A nodes are Down and all network channels are healthy
5. One SC node is Down and one network channel is Down
6. One A node is Down and one network channel is Down
7. One SC and A nodes are Down and one network channel is Down
8. One SC node is Down and two network channels are Down
9. One A node is Down and two network channels are Down
10. One SC and A nodes are Down and two network channels are Down

System response-time distribution, under two cases are shown in Figs. 6.24 and 6.25.

Table 6.8 Probability and hazard rates for Case I

System state	Deadline (ms)	Probability	λ_1^T (h)$^{-1}$	λ_2^T (h)$^{-1}$	λ_3^T (h)$^{-1}$
1.1	30	1.00000	0.0000E+00	0.0000E+00	0.0000E+00
	25	0.99999	3.1728E+00	2.7963E-05	2.4645E-10
1.2	30	0.99990	3.6000E+01	3.5996E-03	3.5996E-07
	25	0.99699	1.0836E+03	3.2519E+00	9.7880E-03
1.3	30	1.00000	0.0000E+00	0.0000E+00	0.0000E+00
	25	0.99996	1.4400E+01	5.7598E-04	2.3039E-08
1.4	30	0.99972	1.0080E+02	2.8216E-02	7.9005E-06
	25	0.99191	2.9124E+03	2.3374E+01	1.8907E-01

First column gives the system state, second time-deadline, third gives corresponding probability of meeting time-deadline in a cycle, columns 4th, 5th and 6th give timeliness hazard rates for $n = 1$, 2 and 3, respectively

Table 6.9 Probability and hazard rates for Case II

System state	Deadline (ms)	Probability	λ_1^T (h)$^{-1}$	λ_2^T (h)$^{-1}$	λ_3^T (h)$^{-1}$
1.1	30	0.99998	7.2000E+00	1.4400E-04	2.8799E-09
	25	0.99978	7.9200E+01	1.7420E-02	3.8324E-06
1.2	30	0.99536	1.6704E+03	7.7150E+00	3.5796E-02
	25	0.98905	3.9420E+03	4.2702E+01	4.6748E-01
1.3	30	0.99998	7.2000E+00	1.4400E-04	2.8799E-09
	25	0.99901	3.5640E+02	3.5249E-01	3.4896E-04
1.4	30	0.99534	1.6776E+03	7.7815E+00	3.6260E-02
	25	0.97441	9.2124E+03	2.3001E+02	5.8785E+00
1.5	30	0.99689	1.1196E+03	3.4712E+00	1.0795E-02
	25	0.97860	7.7040E+03	1.6148E+02	3.4527E+00
1.6	30	0.99999	3.6000E+00	3.6000E-05	3.6000E-10
	25	0.99554	1.6056E+03	7.1293E+00	3.1796E-02
1.7	30	0.99670	1.1880E+03	3.9075E+00	1.2895E-02
	25	0.94573	1.9537E+04	1.0084E+03	5.4444E+01
1.8	30	0.99970	1.0800E+02	3.2390E-02	9.7171E-06
	25	0.93599	2.3044E+04	1.3913E+03	8.8439E+01
1.9	30	0.99997	1.0800E+01	3.2399E-04	9.7197E-09
	25	0.97175	1.0170E+04	2.7962E+02	7.8875E+00
1.10	30	0.99755	8.8200E+02	2.1556E+00	5.2812E-03
	25	0.85557	5.1995E+04	6.6681E+03	9.3560E+02

First column gives the system state, second time-deadline, third gives corresponding probability of meeting time-deadline in a cycle, columns 4th, 5th and 6th give timeliness hazard rates for n = 1, 2 and 3, respectively

Probability that response is generated by 25 and 30 ms, under various operating conditions, along with hazard rate estimated for $n = 1$, 2 and 3 are given in Table 6.8 and 6.9 for case I and II, respectively.

6.6 Summary

Network-induced delays are important for NRT systems as they are the cause of system degradation, failure and sometime system's stability. NRT system's control algorithms considering probabilistic network delay have better control QoP. In this chapter, methods to probabilistically model network-induced delay of two field bus networks, CAN, MIL-STD-1553B and Ethernet is proposed. CAN is random access network. For response-time analysis, various model parameters—probabilities and blocking time pdf- need to be evaluated from message specifications. Effect of hot network redundancy on system delay time of these two networks is analyzed. The method is extended to evaluate sample-to-actuation delay and response-time. A fault-tolerant networked computer system has a number of nodes within sensor, controller and actuator groups, effect to these redundancy on system response-time is also analyzed. Assuming probability of missing deadline in each cycle is constant and given failure criteria, a method to derive timeliness hazard rate is given. This method derives hazard rate from a discrete time process.

References

1. Wesly WC, Chi-Man S, Kin KL (1991) Task response time for real-time distributed systems with resource contentions. IEEE Trans Softw Eng 17(10):1076–1092
2. Diaz JL, Gracia DF, Kim K, Lee C-Gun, Bello LL, Lopez JM, Min SL, and Mirabella O (2002) Stochastic analysis of periodic real-time systems. In: Proceedings of the 23rd IEEE real-time systems symposium (RTSS'02)
3. Diaz JL, Lopez JM, Gracia DF (2002) Probabilistic analysis of the response time in a real time system. In: Proceedings of the 1st CARTS workshop on advanced real-time technologies, October
4. Mitrani I (1985) Response time problems in communication networks. J R Statist Soc (Series B) 47(3):396-406
5. Muppala JK, Varsha M, Trivedi KS, Kulkarni VG (1994) Numerical computation of response-time distributions using stochastic reward nets. Ann Oper Res 48:155–184
6. Trivedi KS, Ramani S, Fricks R (2003) Recent advances in modeling response-time distributions in real-time systems. Proc IEEE 91:1023–1037
7. Muppala JK, Trivedi KS (1991) Real-time systems performance in the presence of failures. IEEE Comp Mag 37–47
8. Sevick KC, Mitrani I (1981) The distribution of queueing network states at input and output instants. J ACM 28(2):353–471
9. Tindell K, Burns A, Wellings AJ (1995) Calculating controller area network (CAN) message response times. Control Eng Prac 3(2):1163–1169
10. Tindell KW, Hansson H, Wellings AJ (1994) Analyzing real-time communications: controller area network (CAN). In: Proceeding of real-time symposium, pp 259–263, December
11. Nolte T, Hansson H, Norstrom C (2002) Minimizing can response-time jitter by message manipulation. In: Proceedings of the 8th real-time and embedded technology and application symposium (RTAS'02)
12. Trivedi KS (1982) Probability & Statistics with Reliability, Queueing, and Computer Science Applications. Prentice-Hall, Englewood Cliffs

13. Nolte T, Hansson H, Norstrom C (2003) Probabilistic worst-case response-time analysis for the controller area network. In: Proceedings of the 9th real-time and embedded technology and application symposium (RTAS'03)
14. Law M, Kelton WD (2000) Simulation Modeling and Analysis. McGraw Hill, New York
15. Nolte T, Hansson H, Norstrom C, Punnekkat S (2001) Using bit-stuffing distributions in can analysis. In: IEEE/IEE real-time embedded systems workshop (RTES'01), December
16. Hansson H, Norstrom C, Punnekkat S (2000) Integrating reliability and timing analysis of can-based systems. In: Proceedings of WCFS'2000-3rd IEEE international workshop on factory communication systems, pp 165–172, September
17. Lindgren M, Hansson H, Norstrom C, Punnekkat S (2000) Deriving reliability estimates of distributed real-time systems by simulation. In: Proceeding of 7th international conference on real-time computing system and applications, pp 279–286
18. Tipsuwan Y, Chow M-Y (2003) Control methodologies in networked control systems. Control Eng Prac 11(10):1099-1111
19. Johnson BW (1989) Design and analysis of fault-tolerant digital systems. Addison Wesley, Reading
20. Cox DR, Miller HD (1970) The theory of stochastic processes. Methuen, London

Chapter 7
Dependability of Networked Computer-Based Systems

7.1 Introduction

A real-time system is said to be operational if it performs its functions correctly and in a timely manner. Performing function correctly is a dependent on healthiness of its constituent components and error free operation of communication links. Ensuring timeliness is dependent upon the delay offered at various stages of node and communication links of the system. So, reliability of a real-time system can be defined as a probabilistic measure of performing correct function and timeliness in the given environment for given amount of time.

We emphasized on a single measure for reliability of real-time systems, which usually is done in two parts. This has following two motivations:

1. real-time system being a single entity so in line with pure hardwired systems it shall have one reliability measure
2. need for a single platform to compare diverse designs, e.g. mainly if one option is either dedicated analog or digital

In this chapter, dependability models—reliability, availability, safety—for NRT system are derived considering timeliness failures in addition to hardware failures.

7.2 Background

The system under discussion is a networked system. A network system has two basic elements, (i) nodes, (ii) network. Nodes are the users of the network and performs the functional part of the system. Network provides a medium for communication between nodes, and responsible for timely behavior of nodes. Network consists of network controller(s), if any, cables, connectors, hub/switches etc. These elements can fail and might have different impact on overall system dependability attributes.

A. K. Verma et al., *Dependability of Networked Computer-based Systems*, Springer Series in Reliability Engineering, DOI: 10.1007/978-0-85729-318-3_7,

Hardware failures consist of failure in nodes and/or network channels. Nodes and networks may have redundancy. Redundant nodes performing similar functions are referred as constituent of a functional group. A functional group will fail if there are failures in more than tolerable number of nodes. A network channel may fail, if it has active components such as switch, controller or hub. Network group is said to be failed if all the network channels are failed.

System state, Π is defined as 4-tuple, $\Pi = (nS, nP, nA, nN)$. Here (nS, nP, nA, nN) is the number of *UP* (or healthy) sensor nodes, processing nodes, actuator nodes and network channels. In each system state there is a finite probability of occurrence of timeliness failure. This probability can be derived from the timeliness hazard rate for that state as derived in Chap. 6.

7.3 Reliability Modeling

Reliability is a well established dependability attribute. In general terms, reliability gives the probability of failure free operation/service by an specified time. For mission-critical and non-repairable systems [8–10], it is the most appropriate dependability attribute. In this section, a method to estimate reliability of networked real-time system is outlined.

7.3.1 System Model

Reliability model for networked real-time systems considered here is based on performability model by Muppala et al. [11]. This model is shown in Fig. 7.1. Muppala [12] has introduced, techniques for the analysis of both hard and soft real-time systems, taking into account the effects of failures and repairs, degraded levels of performance and the violation of response-time deadlines. In this model task arrival are assumed to follow Poisson arrivals, i.e. inter-arrival times are exponentially distributed with rate λ. Also, the task execution times are assumed to be exponentially distributed- with rates α_1 (when 2 processors are working) and α_2 (when 1 processor is working). States labeled 0,1 and 2 represent the number of functioning control systems. The system is considered functional as long as at least one of the processor is operational and no hard deadline has been violated. The parameter c represents coverage probability [13]. State *DF* indicates that the system has failed due to a deadline violation, while *RP* represents failure due to system being in failed state.

When 2 processors are working, two tasks are executed in parallel. The task completion time will be maximum of two task's completion time [14]. The distribution of task completion time, $F_2(t)$ is given as [11]:

$$F_2(t) = \Pr[T_2 \leq t] = 1 - e^{-\alpha_1 t} - e^{-\alpha_2 t} - e^{-(\alpha_1 + \alpha_2)t} \qquad (7.1)$$

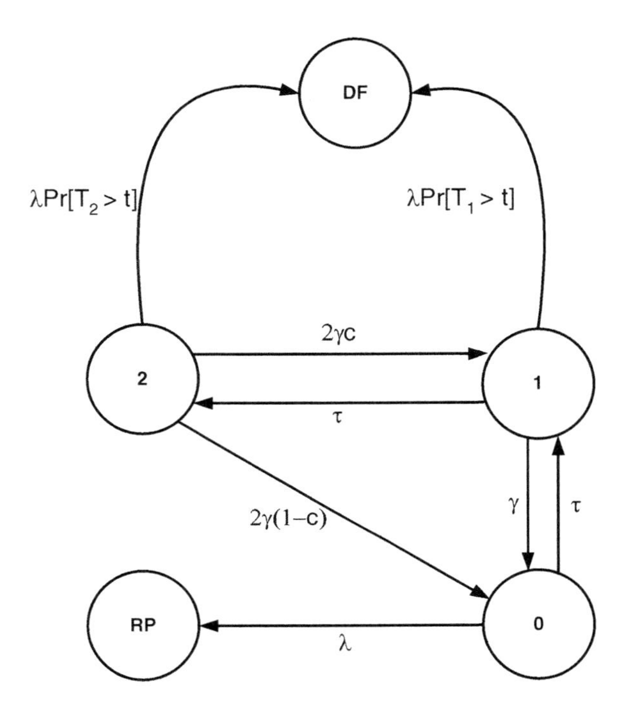

Fig. 7.1 Performability model of Muppala [11]

When only one processor is available, the task has to be executed sequentially, so the task completion time distribution, $F_1(t)$ will be hypoexponential [14] with parameters α_1 and α_2. This is given as [11]:

$$F_1(t) = \Pr[T_1 \le t]$$
$$= \begin{cases} 1 - \dfrac{e^{-\alpha_2}}{\alpha_2 - \alpha_1} e^{-\alpha_1 t} + \dfrac{e^{-\alpha_1}}{\alpha_2 - \alpha_1} e^{-\alpha_2 t}, & \forall \alpha_1 \neq \alpha_2 \\ 1 - e^{-\alpha_1 t} - \alpha_1 t e^{-\alpha_1 t}, & \forall \alpha_1 = \alpha_2 \end{cases} \tag{7.2}$$

The model was solved for system unavailability, due to (i) deadline violations and (ii) due to arrivals when all processors are down.

Reliability model of NRT system proposed here, is conceptually similar to the above model. The key features of proposed model are as follows:

1. tasks arrival are periodic
2. task response-time follow general distribution
3. failure-repair activities at nodes are independent of each other
4. system have shared communication links for information exchange. Shared link have delay which may not be constant.
5. system may have different redundancy configurations

In Chap. 6, it was shown that timeliness failures are dependent on the system state. The reliability of NRT systems is evaluated using the following two step process:

1. Markov model of individual node groups and network
2. From healthy state of NRT system, based on previous step, transitions to failure state because of timeliness hazard rate.

7.3.2 Analysis

Let probability of healthy states of all groups is denoted by $\pi_S^H(t)$, $\pi_C^H(t)$, $\pi_A^H(t)$ and $p_N^H(t)$, where suffix denotes the group, i.e. sensor, controller, actuator and network, respectively. These probabilities are function of time as system evolve with time. These probabilities can be estimated from the Markov model of each group.

With the assumption – failure/repair activities of each group are independent, the NRT system state of healthy states can be obtained from cross product of individual group states [15]. With this system state probabilities can be obtained as product of individual node's state probability. State probability of various system states is given as cartesian cross product of individual group and network state.

$$\pi_{i,j,k,l}^H(t) \in \left\{ \pi_S^H(t) \times \pi_C^H(t) \times \pi_A^H(t) \times \pi_N^H(t) \right\} \tag{7.3}$$

Let timeliness hazard rate from various healthy state of the system is denoted as $\lambda_{i,j,k,l}^T$, where i, j, k, l denote the number of *UP* sensor, controller and actuator node and number of *UP* network channels, respectively. Probability of not reaching *Failure* state by time, t due to timeliness hazard rate from healthy system state is the NRT system reliability, $R(t)$.

$$\frac{dF(t)}{\mathrm{d}t} = \sum_{i \in S} \sum_{j \in C} \sum_{k \in A} \sum_{l \in N} \lambda_{i,j,k,l}^T \pi_{i,j,k,l}^H(t) \tag{7.4}$$

$$R(t) = 1 - F(t) \tag{7.5}$$

7.3.2.1 Example 1

To illustrate the model an example system with two node groups- sensor (controller node is clubbed with sensor node) and actuator are considered. Each node group has 2*oo*3 redundancy. Figure 7.2 shows the node's state-space considering hardware failures. States with more than tolerable number of node failures, i.e. 2, in this case are termed as node failure states. Nodes cannot be repaired back from any of the failure state, while in case of repairable systems, nodes can be repaired from healthy set of system states. In Fig. 7.2 repair rate, μ will be zero in case of non-repairable mission-critical systems.

In this example, there are 2 types of functional nodes and network is assumed to be failure free. So, NRT system will have 4 *UP* states. From the technique

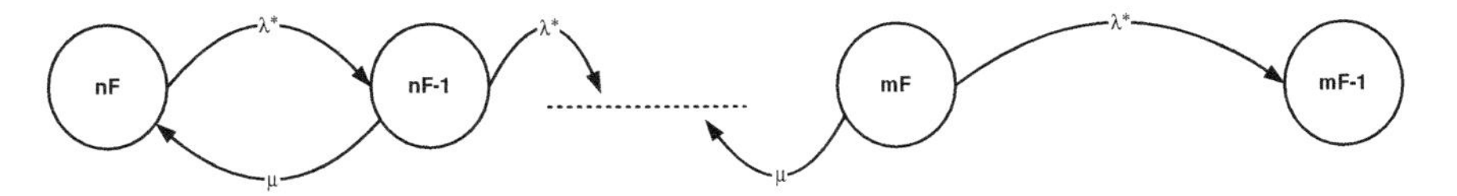

(a) State-space of node groups (S, C and A). nF denotes the total number of UP nodes at beginning. mF denotes the number of nodes required to be in UP state for functional group to be UP. From healthy states repair may be possible. From failure state repair is not possible.

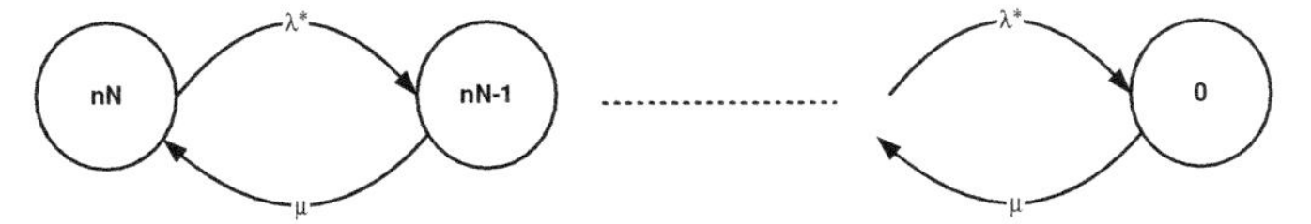

(b) State-space of network. nN denotes the total number of UP network channels at beginning. Repair of network channel may be possible

Fig. 7.2 Generic Markov models for node groups and networks

developed in Chap. 6, timeliness hazard rate for each state can be obtained. Once these hazard rate are available from each of these non-failure state, a state transition diagram for timeliness failure is as shown as in Fig. 7.3.

7.4 Safety Modeling

Safety-critical systems differ from other computer based systems - control and monitoring - based on the mode of operation. Other computer based systems may require change their type of response continuously. While safety systems need to be in either of two states, (i) operate (i.e. allowing EUC to operate), (ii) and safe (i.e. shutting down/stopping of EUC). Means, in absence of any of the safety condition, safety system allows EUC to be in operational state and on assertion of any of the safety condition, safety system takes the EUC in safe state. So, in case of detectable failures, safety system shall take *fail-safe* action.

Safety-critical systems are designed to minimize the probability of unsafe failures, by incorporating design principles such as *fail-safe* and *testability*. To derive safety model for safety critical NRT systems, following assumptions are made.

7.4.1 Assumptions

1. All nodes have indulgent protocol. Indulgent protocol ensures safety even when message arrive late or corrupted [16–18].
2. Safe (unsafe) failure of any group lead to safe (unsafe) failure of system.
3. When system is in safe state, unsafe failure cannot happen.
4. Proof-tests are carried out at system level as a whole.

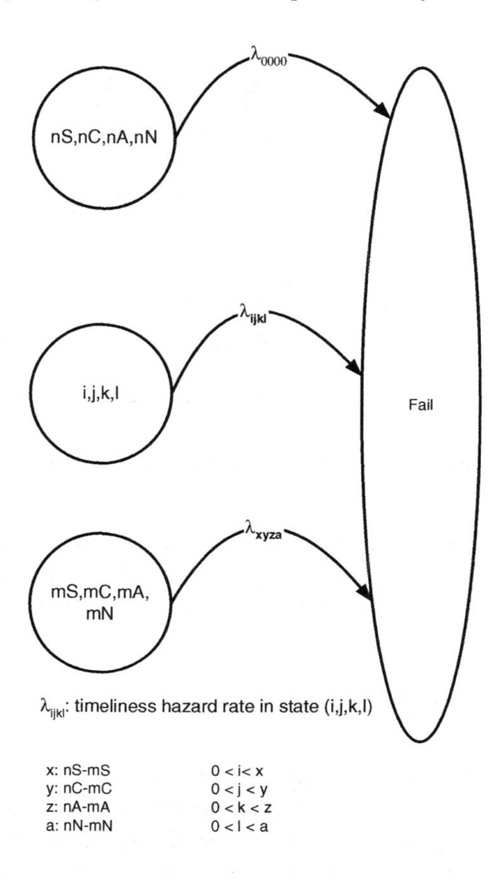

Fig. 7.3 Reliability Model considering timeliness failure

7.4.2 System Model

Safety model of individual nodes without considering timeliness failure is similar to safety model discussed earlier. Failure of network channels, more than tolerable will affect the system, as it will lead to safe failure due to indulgent protocol. Generic safety model of nodes and state-space of network is shown in Fig. 7.4.

State of system safety model can be obtained from the cross product of individual functional group states and network states. From this cross product, functional group safe state and network *DN* state is excluded. As, any node group in safe state or network in *DN* state ensures *fail−safe* failure of the system. This will give exhaustive state-space of system. To estimate *PFaD* of NRT system, mainly system's *DU* state are required.

$$\begin{aligned}\pi^{DU}_{i,j,k,l} &= \pi_{S=\{4\}} \times \pi_{C\in\{1,2,4\}} \times \pi_{A\in\{1,2,4\}} \times \pi^{H}_{N} \\ &+ \pi_{S\in\{1,2,4\}} \times \pi_{C=\{4\}} \times \pi_{A\in\{1,2,4\}} \times \pi^{H}_{N} \\ &+ \pi_{S\in\{1,2,4\}} \times \pi_{C\in\{1,2,4\}} \times \pi_{A=\{4\}} \times \pi^{H}_{N}\end{aligned} \tag{7.6}$$

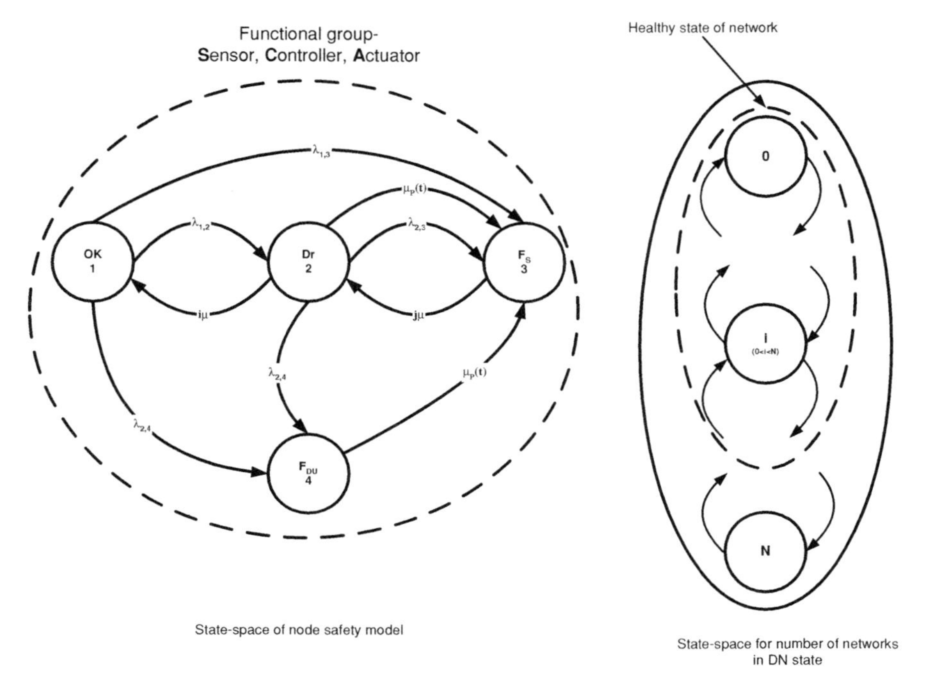

Fig. 7.4 A generic safety model of networked real-time system

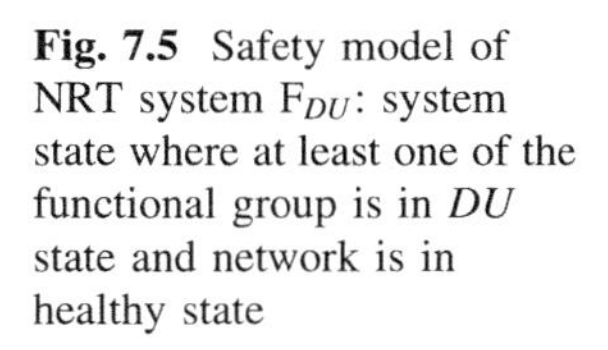

Fig. 7.5 Safety model of NRT system F_{DU}: system state where at least one of the functional group is in *DU* state and network is in healthy state

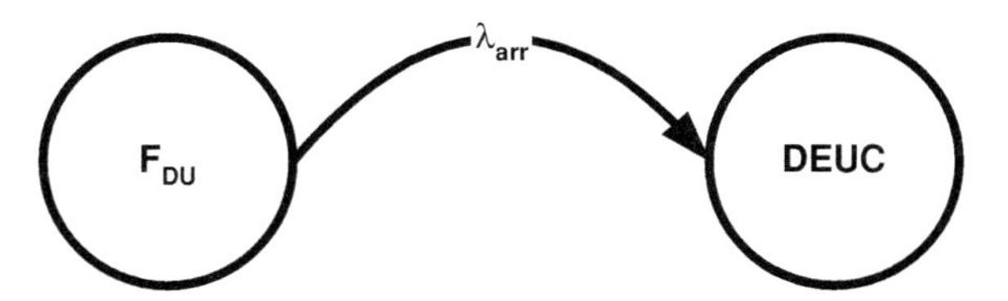

When NRT system is in *DU* state demand arrival will lead to unsafe failure of system. It is pictorially shown in Fig. 7.5.

The probability of NRT system being is *DU* state can be estimated from independent model of functional nodes and network. Once this probability is known $PFaD(t)$ can be estimated.

$$\begin{aligned} \frac{dP_{DEUC}(t)}{\mathrm{d}t} &= \lambda_{\mathrm{arr}} \sum \pi^{DU}_{i,j,k,l}(t) \\ PFaD(t) &= P_{DEUC}(t) \end{aligned} \tag{7.7}$$

In contrast to PES (programmable electronic system), NRT systems have two factors affecting its availability, (i) hardware failures (nodes and networks) and, (ii) timeliness failures. Network failure is due to hardware failure and down time is decided by repair rate. Timeliness failure can occur when system is in *UP* state, down time due to timeliness failure is negligible. So, main contributor to the manifested availability of NRT system are hardware failures. Conditional state probabilities provided system has not failed in unsafe mode by time t.

$$\hat{\pi}_{(i,j,k,l)}(t) = \frac{\pi_{(i,j,k,l)}(t)}{1 - PFaD(t)} \tag{7.8}$$

where
$i \in \{S_1, S_2, S_3, S_4\}$; $j \in \{P_1, P_2, P_3, P_4\}$; $k \in \{A_1, A_2, A_3, A_4\}$; $l \in \{N_H, N_F\}$.

$$mAv(t) = 1 - \sum_{i,j,k,l} \pi_{(i,j,k,l)}(t) \tag{7.9}$$

where $i \in \{S_3\}$; $j \in \{P_3\}$; $k \in \{A_3\}$; $l \in \{N_F\}$.

Timeliness failure of NRT system has negligible effect on its manifested availability, as repair time is negligible. But this momentary safe failure may shutdown the EUC affecting the availability of EUC. Manifested availability in case of PES gives the upper bound of the EUC availability, neglecting other sources causing shutdown of EUC. In case of NRT system manifested availability is unable to give upper bound of the EUC availability.

The upper bound on EUC availability with NRT system can be evaluated from Markov chain of Fig. 7.6.

In Fig. 7.6 transition rate λ^T is the expected timeliness hazard rate of system from *UP* states. Transition rate μ_{EUC} is rate of restoring back EUC to operational state. Computation of expected timeliness hazard rate is given in succeeding section.

7.5 Availability Modeling

A generic Markov model for NRT system availability is shown in Fig. 7.7.

Continuing with philosophy of independence of groups, and evaluation of system state based on individual group states. The availability of NRT system can

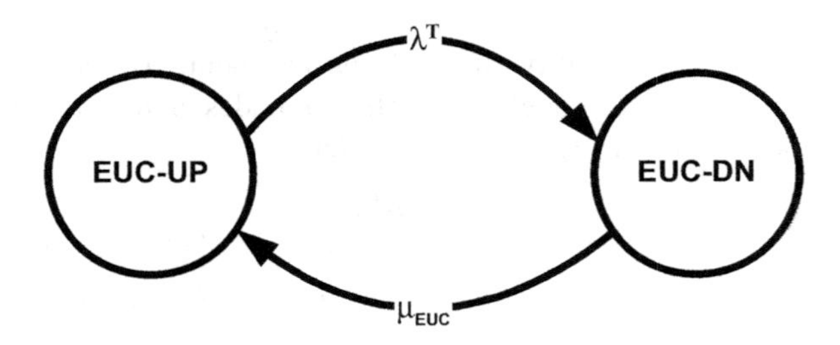

Fig. 7.6 EUC state-space considering timeliness failure when NRT system is in healthy state

be estimated based on the condition that all functional groups are in *UP* state. Failure of any functional group or network lead to unavailability of the system.

So, set of states comprising system availability are those states of functional node and networks, where none is in failure state. Probability of these state can be obtained from individual group's model. Timeliness failures as in case of safety model, have negligible effect on NRT availability, but EUC availability is affected by timeliness failure. Method to find the upper bound of EUC availability is similar as in safety model for NRT systems. So, a new measure to capture probability of timeliness failure is defined.

7.5.1 Timeliness Hazard Rate

Timeliness failures are of importance when NRT system is in *UP* state. For any working state of the system, timeliness failure hazard rate can be evaluated using the technique discussed in preceding chapter. Timeliness hazard rate for a given system state is constant, as derived in previous chapter. Evaluation of equivalent timeliness hazard rate for NRT systems can be modeled as a reward rate problem [19, 20]. Here system evolves because of hardware failures and repairs, timeliness hazard rate is taken as reward rate of the corresponding states.

$$\lambda^T(t) = \frac{\sum_{k \in UP} \lambda_k^T \pi_k(t)}{\sum_{k \in UP} \pi_k} \tag{7.10}$$

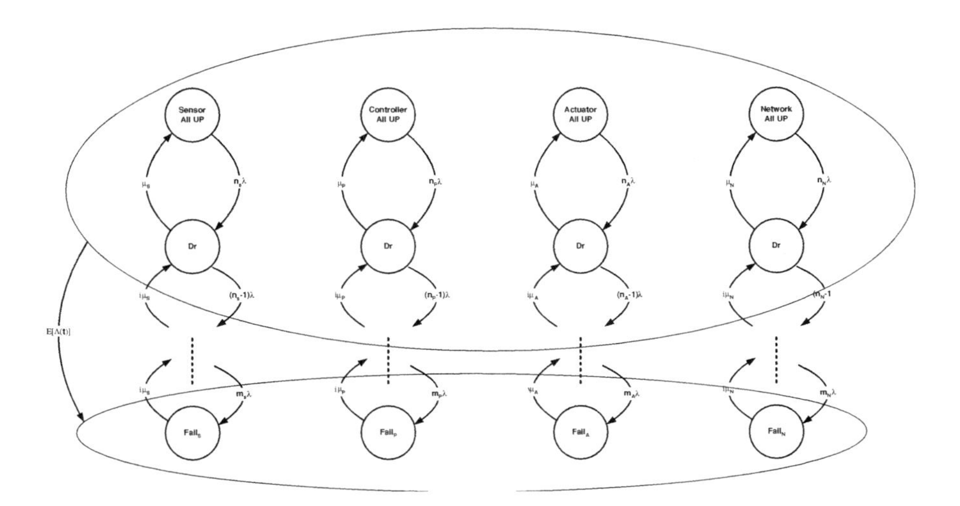

Fig. 7.7 A generic NRT system model for availability

where λ_k is the timeliness hazard rate in system state k; π_k is the probability of being in system state k.

Equation 7.10 gives mean or equivalent timeliness hazard rate at time t. As system state probabilities can be function of time, mean hazard rate over a given time period can be obtained by taking mean over the time period.

$$E\left[\lambda^T(t)\right] = \frac{\int_0^t u.\lambda^T(u)du}{t}$$

In case of availability model, all nodes and networks are repairable, so Markov model of each functional group is irreducible. It implies state probabilities after a long run become independent of time. So, a fixed timeliness hazard rate value can be estimated when system reaches steady state.

7.6 Example

Here example of previous chapter is extended to evaluate reliability, safety and availability. The hazard rate and other parameters of both the nodes is given in Tables 7.1 and 7.2.

Here only CAN network is considered, so probability of network failure is zero. With given system parameters, reliability of the NRT system for a mission time, $t_{\text{mission}} = 4,380\,\text{h}$ is estimate using DSPN based computer tool—TimeNET 4.0 [21]. System is modeled as GSPN [22]. Based on the method developed in this chapter. The GSPN model of NRT system reliability is shown in Fig. 7.8.

Table 7.1 SC and A node hardware failure and repair rates

		SC	A	
Hazard rate (λ)		8.00E-05	4.00E-05	h^{-1}
	Safe (λ_S)	4.00E-05	2.00E-05	h^{-1}
	unsafe (λ_D)	4.00E-05	2.00E-05	h^{-1}
Diagnostic coverage (DC)		0.75	0.75	
Repair rate		1	1	h^{-1}
Diagnostic rate		0.1	0.1	h^{-1}

Table 7.2 Timeliness hazard rate in various system states

System state	n = 2	n = 3	
All node UP	0	0	h^{-1}
1 SC node DN	3.60E-03	3.60E-10	h^{-1}
1 A node DN	0	0	h^{-1}
1 SC and 1 A node DN	2.82E-02	7.90E-06	h^{-1}

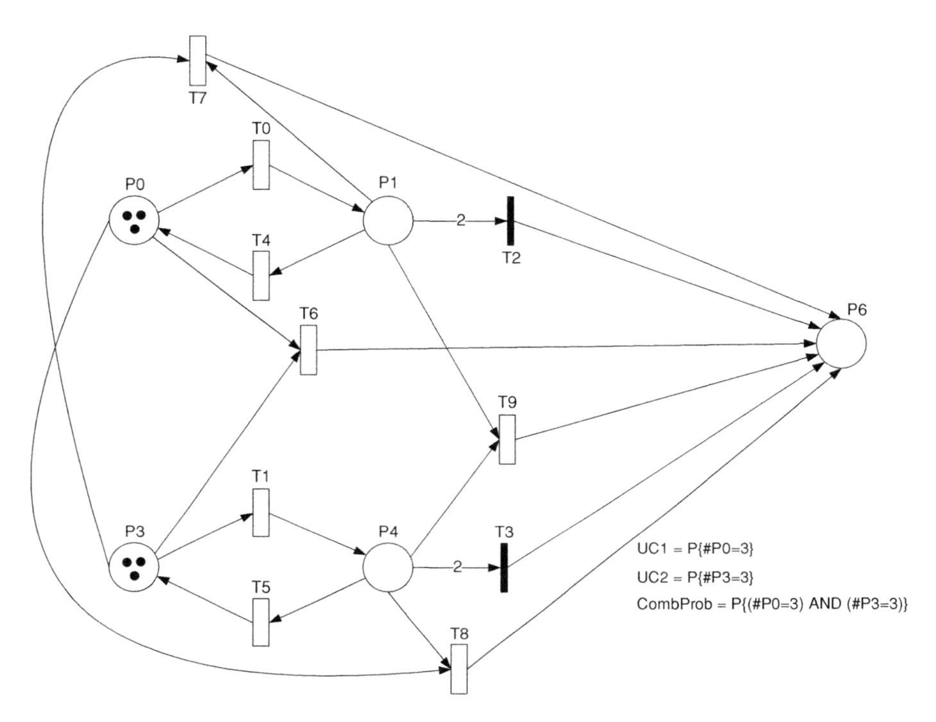

Fig. 7.8 GSPN model of NRT system reliability

Table 7.3 Parameters for safety model

MTBD	4,380 h
T_{proof}	2,000 h
T	8,760 h

Failure distribution evaluated up to $t = 4,380$ h is plotted in Fig. 7.9. Figure 7.9 shows two plots, first one ($n = 2$), is with two consecutive timeliness failures, while second ($n = 3$), is with three consecutive timeliness failures.

For safety modeling, some additional parameters, as MTBD and T_{proof} are need to be specified. These are given in Table 7.3. Here time of operation, $t_{\text{proof}} = 8760$, i.e. one year is taken.

DSPN model based on concept of safety modeling for NRT system is given in Fig. 7.10.

The results are plotted in Fig. 7.11.

GSPN model of the NRT system is shown in Fig. 7.12.

The result of the analysis is summarized in Table 7.4.

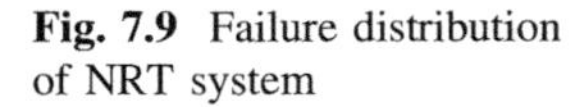

Fig. 7.9 Failure distribution of NRT system

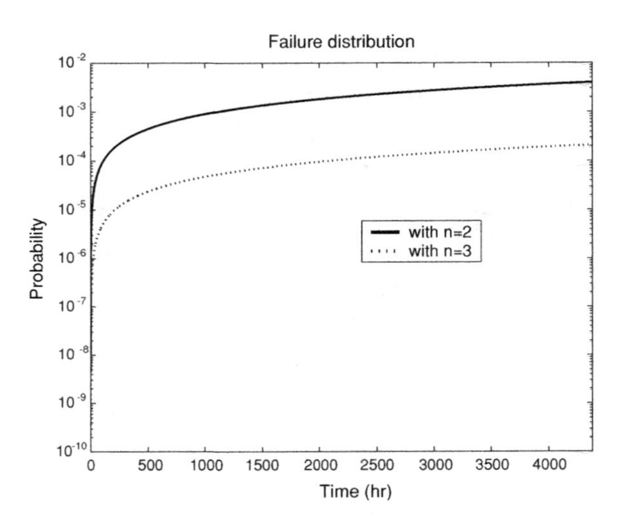

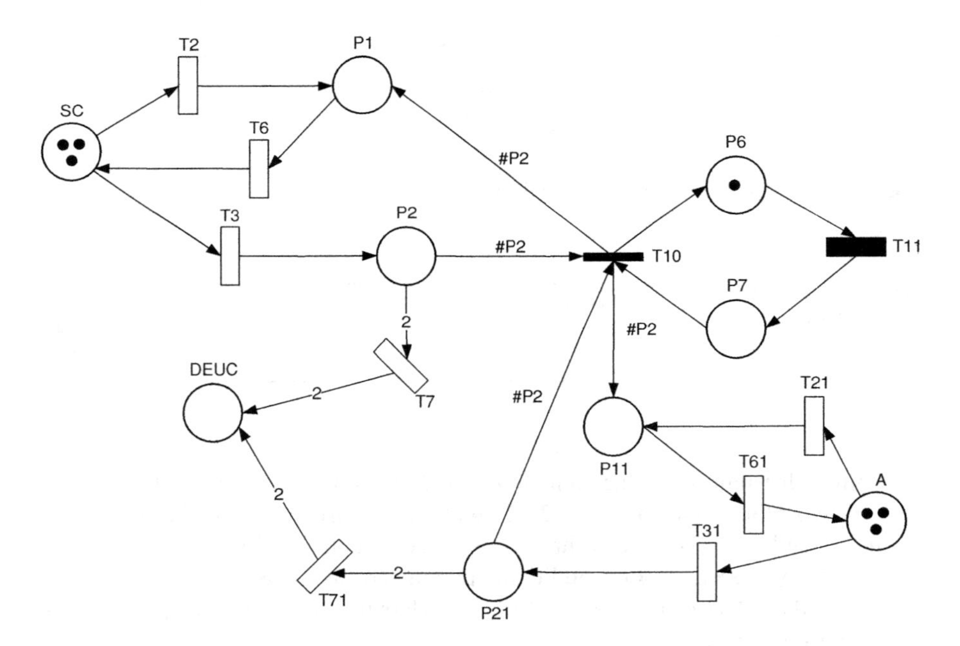

Fig. 7.10 DSPN model of NRT safety

7.7 Summary

Models are derived for three dependability attributes, reliability, availability and safety, of NRT systems. Appropriate engineering assumptions are made about NRT systems used in different applications. It was found as timeliness failure does

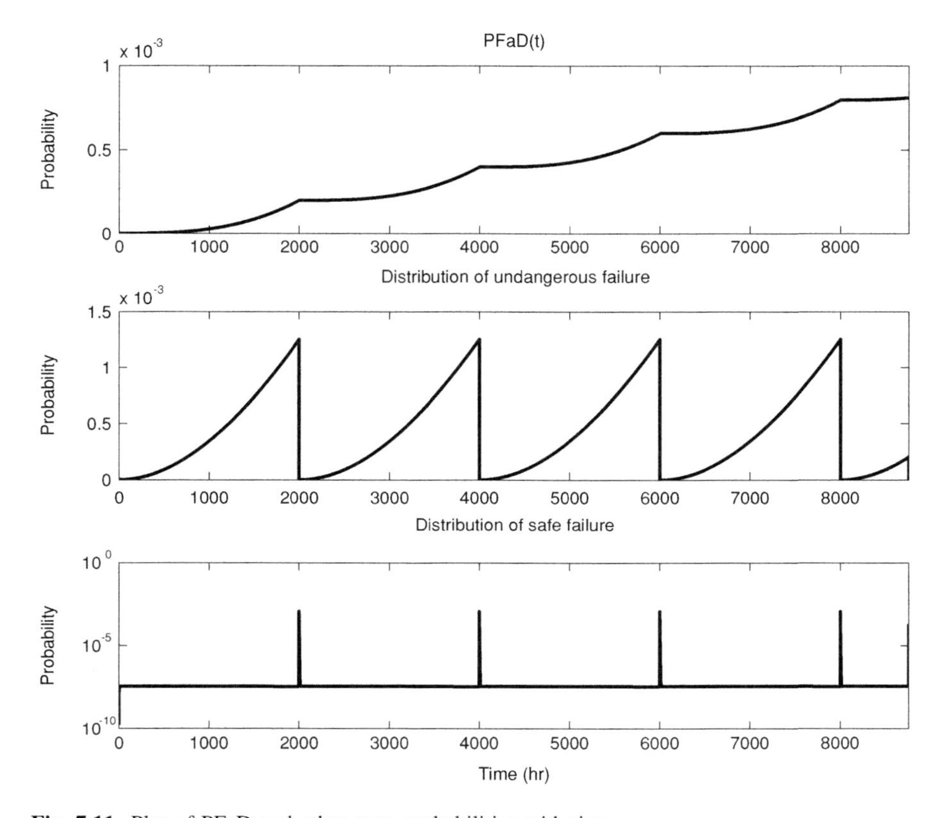

Fig. 7.11 Plot of PFaD and other state probabilities with time

Table 7.4 Mean probabilities of various system states in availability model

System states	Mean probabilities
All healthy	0.9996401
1 SC node DN	0.0002399
1 A node DN	0.00012
1 SC and 1 A node DN	0

not require any repair and if system has not failed can be restarted back instantaneously. In case of safety and availability models, where system goes to *fail–safe* or *unavailable state* for short (or negligible) time, timeliness failures does not affect safety and availability attributes. In these models, timeliness failures only affect the EUC availability. Timeliness hazard rate is modeled as reward rate. This mean reward rate serves as an index for overall timeliness failures. In case of reliability modeling timeliness failure are one source of system failure, they are considered in system reliability modeling.

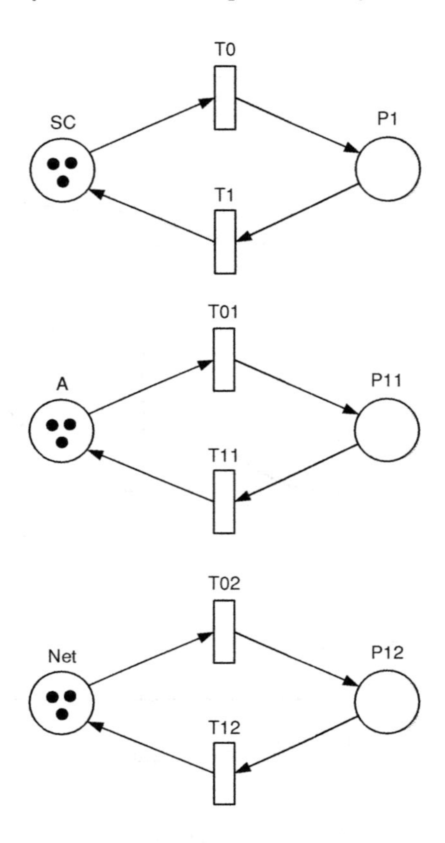

Fig. 7.12 GSPN model for NRT system availability

References

1. Borcsok J, Schwarz MH, Holub P (2006) Principles of safety bus systems. In: UKACC Control Conference, Universities of Glasgow and Strathclyde, UK, September 2006
2. Borcsok J, Ugljesa E, Holub P (2006) Principles of safety bus systems-part II. In: UKACC Control Conference, Universities of Glasgow and Strathclyde, UK, September 2006
3. Elia A, Ferrarini L, Veber C (2006) Analysis of Ethernet-based safe automation networks according to IEC 61508. In: IEEE Conf. on Emerging Technologies and Factory Automation (ETFA '06), pp 333–340, September 2006
4. Rushby J (2001) Bus architectures for safety-critical embedded systems. In Embedded Software, Lecture Notes in Computer Science 2211, Springer, pp 306–323
5. Rushby J (2001) A comparison of bus architectures for safety-critical embedded systems. Technical report, June 2001
6. Vasko DA, Suresh Nair R (2003) CIP Safety: Safety networking for the future. In: 9th International CAN Conference, CAN in Automation (iCC 2003), Munich, Germany
7. IEC 61508: Functional safety of electric/electronic/programmable electronic safety-related systems, Parts 0–7; October 1998–May 2000
8. Avizienis A, Laprie J-C, Randell B (2000) Fundamental concepts of dependability. In: Proc. of 3rd Information Survivability Workshop, pp 7–11, October 2000
9. Johnson BW (1989) Design and analysis of fault-tolerant digital systems. Addison Wesley, New York

10. Mishra KB (1992) Reliability analysis and prediction. Elsevier, Amsterdam
11. Trivedi KS, Ramani S, Fricks R (2003) Recent advances in modeling response-time distributions in real-time systems. In: Proceedings of the IEEE, vol 91, pp 1023–1037
12. Muppala JK, Trivedi KS (1991) Real-time systems performance in the presence of failures. In: IEEE Computer Magazine, pp 37–47, May 1991
13. Dugan JB, Trivedi KS (1989) Coverage modeling for dependability analysis of fault-tolerant systems. IEEE Trans Comput 38(6):775–787
14. Trivedi KS (1982) Probability and statistics with reliability, queueing, and computer science applications. Prentice-Hall, Englewood Cliffs
15. Aldous DJ (1991) Meeting times for independent markov chains. Stoch Process Appl 38:185–193
16. IEC 880: (1986) Software for computers in safety systems of nuclear power stations
17. IEC 60880-2.0: (2006) Nuclear power plants—instrumentation and control systems important to safety—software aspects for computer-based systems performing category a functions
18. Keidar I, Shraer A (2007) How to choose a timing model? In: Proc. 37th Annual IEEE/IFIP Int. Conf. on Dependable Systems and Networks (DSN'07)
19. Howard RA (1971) Dynamic probabilistic systems, vol II: semi-Markov and decision processes. Wiley, New York
20. Muppala J, Gianfranco C, Trivedi KS (1994) Stochastic reward nets for reliability prediction. Commun Reliab Maintainab Serviceability 1(2):9–20
21. Zimmermann A, Knoke M (2007) TimeNET 4.0 user manual. Technical report, August 2007
22. Marson MA, Balbo G, Conte G (1984) A class of generalized stochastic petri nets for the performance evaluation of multiprocessor systems. ACM Trans Comput Syst 93:93–122

Appendix A: MATLAB Codes

MATLAB programs are used throughout the thesis for analysis and plotting of results. Source code of important programs is attached here. The codes are arranged chapter wise.

A.0.0.2 Codes used in Chapter 4

Code for evaluation of safety measure *PFaD* and manifested availability *mAv* for *1oo2* system.

Code: 1

```
% ***********Input Parameters *******************
prmtr = load('parameters1oo20.txt');

LSafe = prmtr(1,1);
LDang = prmtr(2,1);
MeanRepairTime = prmtr(3,1);

DiagCov = prmtr(4,1);
PrTestCov = prmtr(5,1);

Tproof = prmtr(6,1);
RunTime = prmtr(7,1);
MeanTimeBetweenDemands = prmtr(8,1);

CommonCause2 = prmtr(9,1);
CommonCause3 = prmtr(10,1);
NumberofRepairStn = prmtr(11,1);

% **********Derived Parameters ******************
L1 = LSafe + DiagCov*LDang;
L2 = (1-DiagCov)*LDang;
M = 1/MeanRepairTime;
La = 1/MeanTimeBetweenDemands;
a = PrTestCov;
B = CommonCause2;
```

A. K. Verma et al., *Dependability of Networked Computer-based Systems*,
Springer Series in Reliability Engineering, DOI: 10.1007/978-0-85729-318-3,

```
B2 = CommonCause3;
Tp = Tproof;
n = floor(RunTime/Tp);
s = RunTime - n*Tp;
N = NumberofRepairStn;

B2_ = 1 - B2;
B_ = 1 -B;
Alfa = 1-(2-B2)*B;
L = L1 + L2;

% Definition of infinitesimal generator matrix for 1oo1
Q = [   -(2*B_+B)*L        M        0        0        0        0;
         2*B_*L1        -(L+M)      M        0        0        0;
         B*L1             L1       -M        0        0        0;
         2*B_*L2          0         0       -L        M        0;
         0                L2        0        L1      -M        0;
         B*L2             0         0        L2       0      -La;];

% Definition of Delta matrix
Delta = [1 0 0    0     0    0;
         0 1 0    a     0    0;
         0 0 1    0     a    a;
         0 0 0 (1-a)    0    0;
         0 0 0    0  (1-a)   0;
         0 0 0    0     0  (1-a);];


P0 = [1 0 0 0 0 0]';

% *************************************************
E = Delta*expm(Q*Tp);

I = eye(size(Q));

%EPn=(1/RunTime)*(inv(Q)*(expm(Q*s)-I)*inv(I-E)*(I-E^(n+1))*P0);

Temp = inv(Q)*(expm(Q*Tp)-I)*inv(I-E)*(I-E^n)*P0;

meanProb =(1/RunTime)*(Temp + inv(Q)*(expm(Q*s)-I)*P0);

anaPFaD =1 - ones(size(P0'))*(meanProb);

anaFs = [0 1 1 0 1 0]*meanProb;

anaS = [1 0 0 1 0 0]*meanProb;

for i = 0:RunTime/10,
    Time(i+1) = 10*i;
    n = floor(Time(i+1)/Tp);
    s = Time(i+1)-n*Tp;
    Pn(i+1) = (ones(size(P0')))*expm(Q*s)*E^n*P0;
    F(i+1) = 1-Pn(i+1);
```

```
end

% PFaD_t = 1 - ones(size(Pn'))*Pn;
% PFaD = 1 - ones(size(EPn'))*EPn;
% [Tp PFaD]

plot(Time, F);

[anaPFaD anaFs anaS]
Ratio = F(size(F,2))/anaPFaD
%[PFaD_t PFaD]
```

Code for evaluation of safety measure *PFaD* and manifested availability *mAv* for *2oo3* system.

Code: 2

```
% ***********Input Parameters *******************
prmtr = load('parameters2oo3w.txt');

LSafe = prmtr(1,1);
LDang = prmtr(2,1);
MeanRepairTime = prmtr(3,1);

DiagCov = prmtr(4,1);
PrTestCov = prmtr(5,1);

Tproof = prmtr(6,1);
RunTime = prmtr(7,1);
MeanTimeBetweenDemands = prmtr(8,1);

CommonCause2 = prmtr(9,1);
CommonCause3 = prmtr(10,1);

% **********Derived Parameters ******************
L1 = LSafe + DiagCov*LDang;
L2 = (1-DiagCov)*LDang;
M = 1/MeanRepairTime;
La = 1/MeanTimeBetweenDemands;
a = PrTestCov;
B = CommonCause2;
B2 = CommonCause3;
Tp = Tproof;
n = floor(RunTime/Tp);
s = RunTime - n*Tp;

B2_ = 1 - B2;
B_ = 1 -B;
Alfa = 1-(2-B2)*B;
L = L1 + L2;

% Definition of infinitesimal generator matrix for 1oo1
Q = [ -(3*Alfa+3*B2_*B+B2*B)*L      M              0       0
```

```
        0       0       0       0       0       0;
         3*Alfa*L1                      -(2*B_+B)*L-M      M       0
                0       0       0       0       0       0;
         3*B2_*B*L1                     2*B_*L1      -(L+M)      M
                0       0       0       0       0       0;
         B2*B*L1                        B*L1           L1      -M
                0       0       0       0       0       0;
         3*Alfa*L2                        0              0       0
           -(2*B_+B)*L  M       0       0       0       0;
         0                              2*B_*L2          0       0
              2*B_*L1  -(L+M)   M       0       0       0;
         0                                0             L2       0
                B*L1    L1     -M       0       0       0;
         3*B2_*B*L2                       0              0       0
               2*B_*L2  0       0    -(L+La)    M       0;
         0                              B*L2             0       0
                0       L2      0       L1   -(M+La)    0;
         B2*B*L2                          0              0       0
                B*L2    0       0       L2      0     -La;];

% Definition of Delta matrix
Delta = [1 0 0 0    0   0   0   0   0   0;
         0 1 0 0    a   0   0   0   0   0;
         0 0 1 0    0   a   0   a   0   0;
         0 0 0 1    0   0   a   0   a   a;
         0 0 0 0  1-a   0   0   0   0   0;
         0 0 0 0    0 (1-a) 0   0   0   0;
         0 0 0 0   0    0 (1-a) 0   0   0;
         0 0 0 0   0    0   0 (1-a) 0   0;
         0 0 0 0   0    0   0   0 (1-a) 0;
         0 0 0 0   0    0   0   0   0 (1-a);];

P0 = [1 0 0 0 0 0 0 0 0 0]';

% ***************************************************
E = Delta*expm(Q*Tp);

I = eye(size(Q));

% EPn =(1/RunTime)*(inv(Q)*(expm(Q*s)-I)*inv(I-E)*(I-E^(n+1))*P0);

Temp = inv(Q)*(expm(Q*Tp)-I)*inv(I-E)*(I-E^n)*P0;

meanProb =(1/RunTime)*(Temp + inv(Q)*(expm(Q*s)-I)*P0);

anaPFaD =1 - ones(size(P0'))*(meanProb);

anaFs = [0 0 1 1 0 0 1 0 0 0]*meanProb;

anamAv = 1 - (anaPFaD + anaFs);


for i = 0:RunTime/10,
    Time(i+1) = 10*i;
```

```
    n = floor(Time(i+1)/Tp);
    s = Time(i+1)-n*Tp;
    Pn(i+1) = (ones(size(P0')))*expm(Q*s)*E^n*P0;
    F(i+1) = 1-Pn(i+1);
end

% PFaD_t = 1 - ones(size(Pn'))*Pn;
% PFaD = 1 - ones(size(EPn'))*EPn;
% [Tp PFaD]

plot(Time, F);

[anaPFaD anaFs anamAv]
Ratio = F(size(F,2))/anaPFaD
%[PFaD_t PFaD]
```

A.0.0.3 Codes used in Chapter 5

Basic CAN model: The program shown below, is the code for response-time analysis of CAN messages. This program is based on basic CAN model. Message set is defined in file named 'messlist.txt'.

Code: 3

```
% File name: RespCANBasic.m

messList = load('messlist.txt');

messInt = 9;
% Message ID for which response-time distribution is required
ProbDelivery = 0.9999;          %0.99999;
ProbBnB = 0.9999;

bittime = 0.007745;
fractColl = 0.88;

% *********************************************************** %
Util = 0.0;
maxC = 0;

for i = 1: 17
    C(i) = messList(i,2)*8 + 44 +floor((messList(i,2)*8 + 33)/4);
    if i ~= messInt
        Util = Util + (C(i)*bittime)/messList(i,3);
        if C(i) > maxC
            maxC = C(i);
        end
    end
end

% *********************************************************** %
```

```
Pfree = 1 - Util;   %prob. of finding free

% ************************************************************ %

Phigh=1;
for i = messInt+1: 17
    Phigh = Phigh*(1-(fractColl*bittime)/messList(i,3));
    %prob. of no collision

end

% ************************************************************ %

% ******************** (pdf) Blocking time ****************** %
sumTi = 0.0;

for i = 1: 17
    if i ~= messInt
        sumTi = sumTi + 1/messList(i,3);
    end
end

for i = 1: 17
    if i == messInt
        ri(i) = 0.0;
    else
        ri(i) = 1/(messList(i,3)*sumTi);
    end
end

% ********************** Mean Time *************************** %
mT = 0;
for i = 1: 17
    mT = mT + ri(i)*C(i);
end

% *********************************************************** %

temppt = zeros(1,maxC);

for i = 1: 17
    if i ~= messInt
        temppt(C(i)) = temppt(C(i)) + ri(i);
    end
end

%Find non-zero entries
j=1;
for i = 1: maxC
    if temppt(i) > 0
        WaitLen(j)=i;
        WaitVal(j)=temppt(i);
        j=j+1;
    end
```

```
end

pbtfinal = zeros(1,maxC);

for i = 1: length(WaitLen)
pbt(i,:)=WaitVal(i)*(1/WaitLen(i))*
[ones(1,WaitLen(i)),zeros(1,maxC-WaitLen(i))];
pbtfinal = pbtfinal+pbt(i,:);       % pdf of blocking time
end

% ************************************************************ %

% ************* (pdf) Blocking Time by high priority ********** %
sumT = 0.0;
maxCnew = 0;

for i = messInt+1: 17
    sumT = sumT + 1/messList(i,3);
    if C(i) > maxCnew
            maxCnew = C(i);
    end
end

for i = 1: 17
    if i > messInt
        rbhp(i) = 1/(messList(i,3)*sumT);
    else
        rbhp(i) = 0.0;
    end
end

% ********************** Mean Time **************************** %
mTnew = 0;
for i = 1: 17
    mTnew = mTnew + rbhp(i)*C(i);
end

% *********************************************************** %
pbhpt = zeros(1,maxCnew);

for i = 1: 17
    if i > messInt
       pbhpt(C(i)) = pbhpt(C(i)) + rbhp(i);
%         timepbhp(i) = i*bittime;     % ??
    end
end

% ************** Prob[no new hp arrival in Blocking time] ******%
PB_hp = 1;

for i = messInt+1: 17
        PB_hp = PB_hp*(1-(mT*bittime)/messList(i,3));
```

```
        %prob. of no new hp arrival in Blocking
end
% PB_hp = 1-PB_hp;

% ********** Prob[no new hp arrival in Blocking by new time] *****%
PBhp_hp = 1;

for i = messInt+1: 17
        PBhp_hp = PBhp_hp*(1-(mTnew*bittime)/messList(i,3));
             %prob. of no new hp arrival in Blocking by new
end
% PBhp_hp = 1-PBhp_hp;

% ****** (pdf) Blocking in Blocking by new Time by high priority **%
pdfTB = fBlock(PBhp_hp, ProbBnB, pbhpt);

% **************** Cycles required for delivery ****************%
QueueTime = fReady(Phigh, ProbDelivery, pdfTB);

% ********************* Blocking + Queueing Time ************** %
Temp1 = Pfree*QueueTime;
Temp2 = (1-Pfree)*(PB_hp)*conv(pbtfinal,QueueTime);
Temp3 = (1-Pfree)*(1-PB_hp)*conv(conv(pbtfinal,pdfTB),QueueTime);

Temp1=[Temp1,zeros(1,max(length(Temp2),length(Temp3))-length(Temp1))];
Temp2=[Temp2,zeros(1,max(length(Temp2),length(Temp3))-length(Temp2))];
Temp3=[Temp3,zeros(1,max(length(Temp2),length(Temp3))-length(Temp3))];

Bl_QTime = Temp1 + Temp2 + Temp3;

% ********************* Response-time ************************* %
TxTime = zeros(1,C(messInt));
TxTime(C(messInt)) = 1;
Respt = conv(Bl_QTime, TxTime);

CumResp(1)=Respt(1);
Time (1) = bittime;

for i = 2: length(Respt)
    CumResp(i)= CumResp(i-1) + Respt(i);
    Time(i) = i*bittime;
end

plot(Time, CumResp, '-k');
```

Basic CAN response-time model uses two functions to perform the iterative tasks. They are give below:

Function: "fBlock"

```
function [Q] = fBlock(p, C, x)

n = 0;
probSum = 0.0;
```

```
while probSum < C
    n = n + 1;
    prob(n) = ((1-p)^(n-1))*p;
    probSum = probSum + prob(n);
end

prob(n) = prob(n) + 1 - probSum;

convBlock = 1;

mSize = n*length(x)-(n-1);

q = zeros(n,mSize);
Q = zeros(1,mSize);

for i = 1 : n
    convBlock = conv(x,convBlock);
    convBlockt = [convBlock, zeros(1,mSize-length(convBlock))];
    q(i,:) = prob(i)*convBlockt;          %prob(i)*
    Q = Q + q(i,:);
end
```

Function: "fReady"

```
function [Q] = fReady(p, C, x)

n = 0;
probSum = 0.0;

while probSum < C
    n = n + 1;
    prob(n) = ((1-p)^(n-1))*p;
    probSum = probSum + prob(n);
end

prob(n) = prob(n) + 1 - probSum;

convBlock = 1;

mSize = (n-1)*length(x)-(n-2);

q = zeros(n,mSize);
Q = zeros(1,mSize);

for i = 1 : n
    convBlockt = [convBlock, zeros(1,mSize-length(convBlock))];
    q(i,:) = prob(i)*convBlockt;          %prob(i)*
    Q = Q + q(i,:);
    convBlock = conv(x,convBlock);
end
```

For CAN response-time analysis, event based simulation model is as proposed. The code is as follows:

Code: 4

```
clear all;

messList = load('messlist.txt');
bitTime = 0.007745;
RespTime = zeros(17,200);

% Prepare array of time-period and worst-case tx times
for i = 1: length(messList)
    List(i,1)= messList(i,3);
    List(i,2)= (47 + 8*messList(i,2) +
    ceil((34+8*messList(i,2)-1)/4))*bitTime;
end
for run = 1:10000
    % Prepare initial message list
    for i = 1: length(List)
        NextSch(i)= rand*List(i,1);
    end

    RespTime = RespTime + fCANrun(bitTime, List, NextSch);
end
```

CAN response-time simulation model uses one functions to perform the iterative task. It is given below:

Function: "fCANrun"

```
%20.10.2008: saving time changed from 1000 to 2000
function [Resp] = fCANrun(bitTime, List, NextSch)

Resp = zeros(17,200);
Time = 0.0;

while Time < 10000,
% Decide which message to be taken
    % find min in the list
    EarliestTime = min(NextSch);
    if (Time <= EarliestTime)
        %find the message ID
        for i = length(List):-1:1
            if NextSch(i)== EarliestTime
                SelectedMess = i;
                Time = EarliestTime;
                break,
            end
        end
    else
        %search for highest priority to be taken
        for i = length(List):-1:1
            if NextSch(i)<= Time
                SelectedMess = i;
```

```
                break,
            end
        end
    end

    % Update time to completion of Tx
    Time = Time + List(SelectedMess, 2); % + 3*bitTime;

    if Time >= 2000
        rTime = Time - NextSch(SelectedMess);
        Resp(SelectedMess, ceil(rTime*10))=
        Resp(SelectedMess, ceil(rTime*10))+1;
    end

    % Update list
    NextSch(SelectedMess)=NextSch(SelectedMess)+List(SelectedMess,1);

    %Check for missed deadlines
    for i = 1:length(List)
        if NextSch(i) < Time
            if (NextSch(i) + List(i)) < Time
                % deadlines missed, record and update the list
                if Time >= 2000
                    Resp(i,200) = Resp(i,100)+1;
                    %Last element will record the dealline misses
                end
                NextSch(i) = NextSch(i) + List(i);
            end
        end
    end
end
```

Improved CAN response-time model is follows:
Code: 5

```
%Program modified on 04/11/08 to correct inconsistencies
%between previous analytical model's result with that of simulation model.
clear all;

messList = load('messlist.txt');        % load the message file

messInt = 17 ;                           % message of interest
ProbDelivery = 0.9999;                  % 0.99999;

bittime = 0.007745;                     % time in milli-second

% *********************Prob of finding FREE ***************** %
Util = 0.0;
maxC = 0;

for i = 1: 17
    C(i) = messList(i,2)*8 + 44 + floor((messList(i,2)*8 + 33)/4);
    if i ~= messInt
```

```
        Util = Util + (C(i)*bittime)/messList(i,3);
        if C(i) > maxC
            maxC = C(i);
        end
    end
end

Pfree = 1 - Util;   %prob. of finding free

% ************************************************************** %


% ************** (pdf) Blocking time ********************* %
sumTi = 0.0;

for i = 1: 17
    if i ~= messInt
        sumTi = sumTi + 1/messList(i,3);
    end
end

for i = 1: 17
    if i == messInt
        ri(i) = 0.0;
    else
        ri(i) = 1/(messList(i,3)*sumTi);
    end
end

tempPt = zeros(1,maxC);

for i = 1: 17
    if i ~= messInt
        temppt(C(i)) = temppt(C(i)) + ri(i);
    end
end

%Find non-zero entries
j=1;
for i = 1: maxC
    if temppt(i) > 0
        WaitLen(j)=i;
        WaitVal(j)=temppt(i);
        j=j+1;
    end
end

pbtfinal = zeros(1,maxC);

for i = 1: length(WaitLen)
    pbt(i,:)=WaitVal(i)*(1/WaitLen(i))*
    [ones(1,WaitLen(i)),zeros(1,maxC-WaitLen(i))];
    pbtfinal = pbtfinal+pbt(i,:);       % pdf of blocking time
end
```

```
% ************************************************** %


% ********* (pdf) Blocking Time by high priority ************ %


sumT = 0.0;
maxCnew = 0;

for i = messInt+1: 17
    sumT = sumT + 1/messList(i,3);
    if C(i) > maxCnew
            maxCnew = C(i);
    end
end

for i = 1: 17
    if i > messInt
        rbhp(i) = 1/(messList(i,3)*sumT);
    else
        rbhp(i) = 0.0;
    end
end

pbhpt = zeros(1,maxCnew);

for i = 1: 17
    if i > messInt
       pbhpt(C(i)) = pbhpt(C(i)) + rbhp(i);
%         timepbhp(i) = i*bittime;     % ??
    end
end

% ************* Prob[no new hp arrival in Blocking time] *******%
PB_hp = 1;

for i = messInt+1: 17
 PB_hp = PB_hp*(1-(maxC*bittime)/messList(i,3));
 %prob. of no new hp arrival in Blocking
end
% PB_hp = 1-PB_hp;

% ******** Prob[no new hp arrival in Blocking by new time] *****%
PBhp_hp = 1;

for i = messInt+1: 17
        PBhp_hp = PBhp_hp*(1-(maxCnew*bittime)/messList(i,3));
        %prob. of no new hp arrival in Blocking by new
end
% PBhp_hp = 1-PBhp_hp;

% ***** (pdf) Blocking in Blocking by new Time by high priority***%
pdfTB = fBlock(PBhp_hp, ProbDelivery, pbhpt);
```

```
% **************** Cycles required for delivery **************%
% QueueTime = fReady(Phigh, ProbDelivery, pdfTB);

% *********** Blocking + Queueing Time *********************** %
Temp1 = Pfree;                    %*QueueTime;
Temp2 = (1-Pfree)*(PB_hp)*pbtfinal;     %conv(pbtfinal,QueueTime);
Temp3 = (1-Pfree)*(1-PB_hp)*conv(pbtfinal,pdfTB);
%conv(conv(pbtfinal,pdfTB),QueueTime);

Temp1=[Temp1,zeros(1,max(length(Temp2),length(Temp3))-length(Temp1))];
Temp2=[Temp2,zeros(1,max(length(Temp2),length(Temp3))-length(Temp2))];
Temp3=[Temp3,zeros(1,max(length(Temp2),length(Temp3))-length(Temp3))];

Bl_QTime = Temp1 + Temp2 + Temp3;

% **************** Response-time ************************* %
TxTime = zeros(1,C(messInt));
TxTime(C(messInt)) = 1;
Respt = conv(Bl_QTime, TxTime);

CumResp(1)=Respt(1);
Time (1) = bittime;

for i = 2: length(Respt)
    CumResp(i)= CumResp(i-1) + Respt(i);
    Time(i) = i*bittime;
end

plot(Time, CumResp, '-k');
```

Function "fBlock" used here is same as of basic CAN model.
Code for evaluating timeliness hazard rate is as follows:
Code: 6

```
% Hazard rate evaluation based on probability of meeting
% deadline and criteria of timeliness failure, i.e. n=1,2,3,...


p = 0.99998;
q = 1-p;
cycTime = 1; %time in seconds
operTime = 1000; % time in hour

P1 = [p q; 0 1];

P2 = [p q 0; p 0 q; 0 0 1];

P3 = [p q 0 0; p 0 q 0; p 0 0 q; 0 0 0 1];

P10 = [1 0];
P20 = [1 0 0];
P30 = [1 0 0 0];
```

```
for i = 1: 2000,
    Pz1(i,:) = P10*P1^(i);
    Pz2(i,:) = P20*P2^(i);
    Pz3(i,:) = P30*P3^(i);
    if i > 1
        lam1(i-1) = ((Pz1(i,2)-Pz1(i-1,2))/(1-Pz1(i-1,2)));
        lam2(i-1) = ((Pz2(i,3)-Pz2(i-1,3))/(1-Pz2(i-1,3)));
        lam3(i-1) = ((Pz3(i,4)-Pz3(i-1,4))/(1-Pz3(i-1,4)));
    end
end
```

Code for comparison of failure rate from DTMC and exponential failure distribution.

Code: 7

```
% Comparision of failure distribution of DTMC
% and eqvt exp distribution, for n=1,2,3. For 50ms cycle.

p = 0.99998;
q = 1-p;

lam1 = 2e-5;      % Hazard rate in per cycle
lam2 = 4E-10;
lam3 = 8E-15;

cycTime = 50; %time in milliseconds
operTime = 10; % time in hour

P1 = [p q; 0 1];

P2 = [p q 0; p 0 q; 0 0 1];

P3 = [p q 0 0; p 0 q 0; p 0 0 q; 0 0 0 1];

P10 = [1 0];
P20 = [1 0 0];
P30 = [1 0 0 0];

for i = 1:10000,            %operTime*3600000/50,
    Pz1(i,:) = P10*P1^(1000*(i-1));
    Pz2(i,:) = P20*P2^(1000*(i-1));
    Pz3(i,:) = P30*P3^(1000*(i-1));

    Pd1(i) = Pz1(i,2);
    Pd2(i) = Pz2(i,3);
    Pd3(i) = Pz3(i,4);

    Pl1(i) = 1-exp(-(i-1)*1000*lam1);
    Pl2(i) = 1-exp(-(i-1)*1000*lam2);
    Pl3(i) = 1-exp(-(i-1)*1000*lam3);

    Err1(i) = (Pd1(i)-Pl1(i)); %/Pd1(i))*100;
    Err2(i) = (Pd2(i)-Pl2(i)); %/Pd2(i))*100;
```

```
    Err3(i) = (Pd3(i)-Pl3(i)); %/Pd3(i))*100;
end
Time = [0:1:9999]*5;     %time in seconds

plot(Time, Err1, 'k', Time, Err2, 'r', Time, Err3, 'b');
```

9780857293176